TL 9000 Release 3.0: A Guide to Measuring Excellence in Telecommunications

Second Edition

TL 9000 Release 3.0: A Guide to Measuring Excellence in Telecommunications

Second Edition

Sandford Liebesman, Alka Jarvis,
and Ashok V. Dandekar

ASQ Quality Press
Milwaukee, Wisconsin

TL 9000 Release 3.0: A Guide to Measuring Excellence in Telecommunications,
Second Edition
Sandford Liebesman, Alka Jarvis, and Ashok V. Dandekar

Library of Congress Cataloging-in-Publication Data

Liebesman, Sandford, 1935–
TL 9000 release 3.0: a guide to measuring excellence in telecommunications / Sandford Liebesman, Alka Jarvis, and Ashok V. Dandekar.—2nd ed.
p. cm.
Rev. ed. of: TL 9000. c2001.
Includes bibliographical references and index.
ISBN 0-87389-542-8
1. Telecommunication—Quality control. 2. TL 9000 (Standard) I. Jarvis, Alka. II. Dandekar, Ashok V., 1951– III. Liebesman, Sandford, 1935–. TL 9000. IV. Title.

TK5102.84 .L55 2002
384'.068—dc21 2002003681

10 9 8 7 6 5 4 3 2 1

ISBN 0-87389-542-8

Acquisitions Editor: Annemieke Koudstaal
Project Editor: Craig S. Powell
Production Administrator: Gretchen Trautman
Special Marketing Representative: David Luth

Printed in the United States of America

Printed on acid-free paper

American Society for Quality

ASQ

Quality Press
600 N. Plankinton Avenue
Milwaukee, Wisconsin 53203
Call toll free 800-248-1946
Fax 414-272-1734
www.asq.org
http://qualitypress.asq.org
http://standardsgroup.asq.org
E-mail: authors@asq.org

Dedication

We dedicate this book to Henry Malec, a colleague and good friend. Henry joined the QuEST Forum at the start and contributed to the development of the Requirements and Measurements (Metrics) Handbooks. He was the principal creator of the Measurements (Metrics) Repository System, in which he used a double blind process to assure the integrity of the data. We relied on Henry to give us a clear view of where he thought the QuEST Forum should be heading. Sadly, he was taken from us much too early.

Table of Contents

Foreword

The telecommunications industry is globalizing, restructuring, and growing dramatically to become the backbone of the new digital economy. Some of the world's largest companies are in a headlong competition to provide a high-speed digital world connection to customers, either by telephone lines, cable TV connections, wireless, or satellite connections. The quality and reliability of these new digital networks, and the supply lines required to build them, must continually improve to deliver on the full promise of the new technology on a global basis. The QuEST Forum, a collaborative effort between telecommunications service providers and their suppliers, addressed this challenge by developing and deploying the TL 9000 Quality Management System focused on improving the performance of the new digital telecom industry.

TL 9000 is about improving performance: better overall product quality, lower costs, reduced cycle time, and improved customer satisfaction. The requirements and measurements build on currently used industry standards, including the ISO 9000 series, and are rapidly gaining worldwide momentum and acceptance. They're specific to Telecom, which creates the opportunity to measure and drive industry improvement.

This book captures and interprets the TL 9000 Quality Management System requirements, and cost- and performance-based measurements. It provides a step-by-step approach by first discussing the standard, describing the rationale for the existence of the standard, and then sharing tips on how to meet the requirements. The authors have been participants in the QuEST Forum from its inception and have worked diligently to write this book. They show us that the solution for better quality is not just testing, inspections, or documented procedures but requires a broad and realistic view of customer requirements and how products and services can affect these requirements. We are reminded that successful processes are vital to the reputation and health of an organization.

As you read this book, be prepared to be challenged by new concepts that can produce improved software, hardware, service quality, and satisfied and loyal customers. It will be an important tool to aid in the improvement of telecommunications products and services in any company choosing to use it.

Stephen G. Welch
President
SBC Procurement/Corporate Real Estate
QuEST Forum Founding Chairman

Preface

The Quality Excellence for Suppliers of Telecommunications (QuEST) Leadership Forum was formed as a telecommunications sector consortium in January 1998. The QuEST Forum launched a worldwide supply-chain initiative to develop quality management system standards aimed at improving the products and services provided by the industry. The dynamic nature of telecommunications and the globalization of the industry highlighted a need for a single set of telecommunications quality management system standards. The Forum brought together service providers and suppliers early in 1998 and produced the first two handbooks, Release 2.5 of *TL 9000 Quality System Requirements*[1] and *TL 9000 Quality System Metrics (Measurements)*,[2] by the end of 1999. The Forum also decided to upgrade the handbooks when ANSI/ISO/ASQ Q9001-2000 was published in December 2000. Release 3.0 of each was published in March 2001,[3,4] within three months of the publication of ANSI/ISO/ASQ Q9001-2000.

The industry makes annual purchases estimated at more than $125 billion worldwide. Of this, the cost of poor quality is estimated to be $10–15 billion. The potential savings from implementing TL 9000 is therefore considerable. This is especially important since it is estimated that the standards will be applicable to over 10,000 suppliers worldwide.

One major new aspect of TL 9000 is the measurements* defined for hardware, software, and service quality. It is expected that the measurements will be a major benchmarking tool that will help the industry improve quality and, hence, drive down cost. Historically the industry has benefited from hardware quality improvement. Between 1985 and 1996, the annual cost of poor quality (COPQ) of hardware within the U.S. telecommunications industry was measured at $2.5 billion; but as the products improved, the annual COPQ was reduced to $750 million.

The three authors of this book have worked on TL 9000 from the beginning and bring to the table strong backgrounds in the hardware, software, and service aspects of telecommunications. They are also well versed in the issues debated during the development of TL 9000, the industry need to develop additional requirements beyond ISO 9001,[5] and the identification of measurements that are aimed at

*The original terminology used was "metrics." In 2000, this was changed to "measurements." We will use these terms interchangeably throughout the book.

driving continual improvement. They also have experience implementing TL 9000 in their organizations.

The Forum started with 44 member companies of the telecommunications industry. As of December 31, 2001, there were 155 members, including 14 service providers, 92 suppliers, and 49 liaison members.

This book presents information in seven chapters. The introductory chapter covers the creation of the QuEST Forum, its membership categories, the two handbooks, key differences between ANSI/ISO/ASQ Q9001-2000 and TL 9000, the registration process, and the QuEST Forum's global growth initiative. Chapter 2 describes the additional common requirements (adders) of TL 9000. The adders are highlighted and explained, and "tips" on implementation are provided. Chapter 3 covers the same material for the hardware, software, and service adders. Chapter 4 covers the individual measurements and describes their purposes. Chapter 5 describes the measurements collection process, analysis of the data, and the product categories. Chapter 6 discusses the case studies and data received from the pilot projects. Finally, chapter 7 concludes with the benefits of TL 9000 implementation and future activities of the QuEST Forum. The appendixes provide a table of adders in terms of ANSI/ISO/ASQC Q9001-1994[6] and ANSI/ISO/ASQ Q9001-2000 and a description of the accreditation/registration process. Each chapter contains a list of references. A bibliography is provided at the end of the book.

ENDNOTES

1. The QuEST Forum, *TL 9000 Quality System Requirements*, Book One, Release 2.5 (Milwaukee: ASQ Quality Press, 1999).
2. The QuEST Forum, *TL 9000 Quality System Metrics*, Book Two, Release 2.5 (Milwaukee: ASQ Quality Press, 1999).
3. The QuEST Forum, *TL 9000 Quality Management System Requirements Handbook*, Release 3.0 (Milwaukee: ASQ Quality Press, 2001).
4. The QuEST Forum, *TL 9000 Quality Management System Measurements Handbook*, Release 3.0 (Milwaukee: ASQ Quality Press, 2001).
5. ANSI/ISO/ASQ Q9001-2000, *Quality Management Standards—Requirements,* 3rd ed. (Milwaukee: ASQ Quality Press, 2000).
6. ANSI/ISO/ASQC Q9001-1994, *Quality Systems—Model for Quality Assurance in Design, Development, Production, Installation and Servicing*, 2nd ed. (Milwaukee: ASQ Quality Press, 1994).

Acknowledgments

The authors would like to thank all the past and present members of the QuEST Forum whose support made it possible for us to write this book. We would especially like to thank Steve Welch, President, SBC Procurement/Corporate Real Estate, for his support and the Foreword to this book; Jim Mroz of The Informed Outlook, for contributing to the development of chapter 6 on TL 9000 case studies; and Dawn Perry of Cisco Systems for contributing the appendix on accreditation and registration.

We would like to thank our editor, Annemieke Koudstaal, for her support and counsel during the development of the second edition of the text. In addition, we would like to acknowledge the support of the following individuals:

Name	Company	Support
Reg Blake	BSI	Pilot Survey
Bob Brigham	Telcordia Technologies	Seed documents and planning the structure of the book
Kevin Calhoun and Sandy Holston	Siecor	Pilot Survey
Don Carlson and Art Morrical	Lucent Technologies	Pilot Survey
Michael Faccone	Telcordia	Pilot Survey
Rolf Hendrickson	Nortel Networks	Pilot Survey
Wally Juban	Pirelli Cables	Pilot Survey
Don Moore	Fujitsu	Pilot Survey
Malcolm Phipps	QMI	Pilot Survey
Gary Reams	NEC America	Pilot Survey
Donna Reinsch	Marconi	Pilot Survey
Pierre Salle	KEMA	Pilot Survey
Joel Savilonis	NSAI	Pilot Survey

Name	Company	Support
Joe Taylor	Tellabs	Pilot Survey
Jim Woodard	SBC–CA	Pilot Survey
Rose Hoff	Advanced Fiber Communications (AFC)	Reviewer
Rich Morrow	UTD	Reviewer
Zvi Ravia	Comverse	Reviewer
John Walz	Lucent Technologies	Reviewer
Rick Werth	SBC	Reviewer
Tom Yohe	Alcatel	Reviewer

1

Introduction

CREATION OF THE QUEST FORUM

In early 1996, Bell Atlantic, BellSouth, Pacific Bell (later acquired by SBC), and Southwestern Bell (SBC) executives initiated an effort to establish better quality management requirements for the industry. They represented the future expectations of the leading telecommunications service providers in the United States.

After a year and a half of planning, the four service providers invited their leading suppliers to a meeting in Baltimore in October 1997. The result was the formation of a cooperative venture among the service providers and suppliers to create an industry quality management system standard aimed at improving the quality of telecommunications products and services.

Work began in January 1998 to draft a set of quality requirements to be incorporated into a TL 9000 handbook. To accomplish this, work groups consisting of industry experts in software, hardware, and services were formed. The work groups were chartered with the development of the TL 9000 hardware, software, and service requirements to be added to the ISO 9001[1] requirements. These added requirements, or "adders," would be based on the historical needs of the industry. Additional work groups were formed to develop a second handbook consisting of hardware, software, and service metrics (measurements). A regular monthly schedule of team meetings was initiated with the goal of having a working implementation program by the end of 1999.

It was understood from the beginning that cost- and performance-based metrics (measurements) were especially important to the effort. This is because they can be used to quantify the benefits gained, assess progress in quality maturity, and identify areas where the quality process improvement will have the greatest cost impact and provide comparative benchmarking capabilities for the industry.

The QuEST Forum recognized the parallel between what they wanted to accomplish and what had been done by the automotive industry a few years earlier. The auto industry had created a standard based on ISO 9001 with adders called QS-9000.[2] The Forum consulted with the auto industry forum, the Automotive Industry Action Group (AIAG), during the start-up phase.

The Forum also recognized the need to remain current with international quality management system standards being developed by ISO/TC 176. Since the Forum is a liaison member of the technical committee, the Requirements Working Group received early and continuing information on the restructuring of ISO 9001. Thus, when the revised ISO 9001 standard was published in December 2000, the working group had developed a compatible requirements handbook that was published in March 2001. At the same time, the term *metrics* was changed to *measurements* and a revised measurements handbook was published.

TELECOMMUNICATIONS INDUSTRY NEEDS

The working groups reviewed ISO 9001 and determined that supplemental requirements were needed in the following areas:

- *Reliability and associated costs.* The industry places a great emphasis on reliability because of end-user demands placed on service providers. The networks, switching systems, and local transmission equipment must operate continuously to accurately pass information between end users. In this high technology information age, the rapid transfer of data requires highly reliable products and services.

- *Software development and lifecycle management.* Software is an integral part of telecommunications product and service offerings. Although ANSI/ISO/ASQ Q9000-3-1997 is a guide for applying ANSI/ISO/ASQC Q9001-1994 to software quality management systems, it does not cover the software lifecycle needs of the industry. TL 9000 developers used inputs from ANSI/ISO/ASQ Q9000-3-1997,[3] ISO/IEC 12207,[4] TR-NWT-000179[5], and other sources to satisfy most of these needs.

- *Specialized service functions.* Installation, engineering, maintenance repair, and customer support services are integral to the customer–supplier relationship in the telecommunications industry. The coverage of post-delivery customer support services in clauses 7.2.1 and 7.5.1 in ANSI/ISO/ASQC Q9001-2000[1] is extremely limited and additional requirements are needed to provide support functions such as emergency service, problem notification, and supplier support.
- *Continuation and further development of the relationships between service providers and suppliers.* A unique characteristic of the telecommunications industry is the ongoing relationship between service providers and their suppliers. Service providers require new or improved features on a regular basis. Thus there is a need for continual communication between customers and their suppliers.

These industry needs led to the development of the 81 adders and the 11 measurements defined in the two TL 9000 handbooks.

THE GOALS AND EXPECTATIONS OF TL 9000

There were five major goals defined by the QuEST Forum:

1. To foster systems that will protect the integrity of telecommunications products, services, and networks
2. To develop requirements that will help organizations more accurately assess the implementation of their quality management system
3. To identify tools that will drive continual improvement of products, services, and processes
4. To develop standard measurements for use as a continual improvement tool
5. To implement an industry standard assessment process that will reduce the multiplicity of conflicting programs

There are a number of industry benefits that the QuEST Forum membership expects from the development of TL 9000. These include:[6]

- Improved service to end users
- Enhanced customer–supplier relationships (Note: The QuEST Forum was designed as a cooperative effort between customers and suppliers)

- A reduction in the number of external 2nd and 3rd party audits and site visits
- Uniform performance- and cost-based measurements for use as benchmarks for improving product and service quality
- Enhanced supply chain management, including second- and third-tier suppliers
- The creation of a platform for industry improvement initiatives

Specific expectations for Forum member products and services include:

- Cycle-time reduction
- Improved on-time delivery
- Reduced lifecycle costs
- Superior products
- Defect reduction
- Increased profitability and market share

QuEST Forum Membership Categories

There are four classes of membership in the QuEST Forum:

1. Service providers
2. Suppliers
3. Liaison members
4. Nonmember liaisons

A fifth class, recognized consultants, is under consideration.

Service providers and suppliers are full members with full privileges and responsibilities. The privileges include voting rights, access to data, and full participation in working groups. The responsibilities include full payment of dues (currently $10,000 per year), a requirement to provide at least one subject matter expert for the working groups, and attendance at a specified number of meetings.

Liaison members come from a number of communities such as the accreditation agencies, registrars, training organizations, consultants, other standards bodies, and other industry organizations. They do not have voting privileges or access to the data and may be restricted from membership in certain working groups. Their responsibilities include a reduced payment of dues (currently $2000), and a prohibition against commercial activities while attending official Forum meetings. Each liaison member must be sponsored by at least two full members.

Nonmember liaisons are organizations that have a specific expertise or represent a body of knowledge and/or are organizations that provide a resource critical to the Forum's future growth and expansion of the TL 9000 programs. There are no membership fees and they are limited to specific activities that utilize their expertise. An example of a nonmember liaison is the International Accreditation Forum.

Recognized consultants are under consideration as a membership class. The proposal is that to be recognized, they must attend a QuEST Forum–sanctioned TL 9000 course, pass a written exam, and attend at least one "Consultant Information Exchange" session each year. They would also have to agree that their TL 9000 clients complete a Forum administered Customer Satisfaction Survey, adhere to the QuEST Forum Representatives Guiding Principles, and agree to resolve any conflicts or problems by using the QuEST Forum mediation board. Consultants would have to be sponsored by at least two full members.

The major benefits for both the full and liaison members are:

- *Participation in the development of the quality requirements.* A company that joins the Forum as a full or liaison member has access to and helps develop TL 9000 requirements and measurements handbooks. The results for member companies should be:
  - Increased efficiency and productivity as a result of their efforts in developing the TL 9000 handbooks.
  - More uniform internal and external audits.
- *Voting rights.* Full member: one vote per member company on all decisions. Liaison members cannot vote.
  - Thus, member companies play a part in developing consensus quality system requirements, measurements, and bylaws.
- *Industry data for all product categories.* Viewing privileges for all of the aggregate industry data.
  - Full members get free access to valuable benchmarking industry data for use in comparing and improving products and services.
  - Liaison members do not have access to the data, but they may purchase the data.

- *Work group interaction.* Full members must provide at least one subject matter expert to participate in work group activities. However, they may provide as many subject matter experts to work on as many work groups as they desire.
  - Members receive valuable information and understanding of requirements and measurements and have a voice in shaping them.
- *Forum meetings held three or four times per year.* Members can send representatives to Forum meetings.
  - Members hear informative speakers, receive updates from the Executive Board, participate in workshops, and establish key networking contacts.
- *Forum member discounts.*
  - Members receive discounts on Forum products such as the handbooks, Forum merchandise, and training materials. Site licenses have been developed for the training courses and for obtaining copies of the handbooks at reduced prices.

The TL 9000 Model

The basic foundation of the TL 9000 model is ANSI/ISO/ASQ Q9001-2000. TL 9000 incorporates the requirements of that standard and adds 81 new requirements, or "adders." The adders are categorized as common requirements, hardware-specific requirements, software-specific requirements, service-specific requirements, or paired combinations of specific requirements.

A second set of requirements covers the measurements. Again, these are categorized into common, hardware-specific, software-specific, and service-specific measurements.

Figure 1.1 depicts the TL 9000 model.

The TL 9000 Handbooks

The Forum developed Release 3.0 of the handbooks during 2000. The *Requirements Handbook*[6] contains the verbatim ISO 9001 requirements, the 81 adders, and appendixes covering accreditation, registration, and guidance for customer–supplier communication. The *Measurements Handbook*[7] contains a description of the measurements, responsibilities of all parties with respect to the measurements, the product category and normalization tables, and customer satisfaction measurements guidelines. Updated product category tables can be found on the QuEST Forum Web site (www.questforum.org).

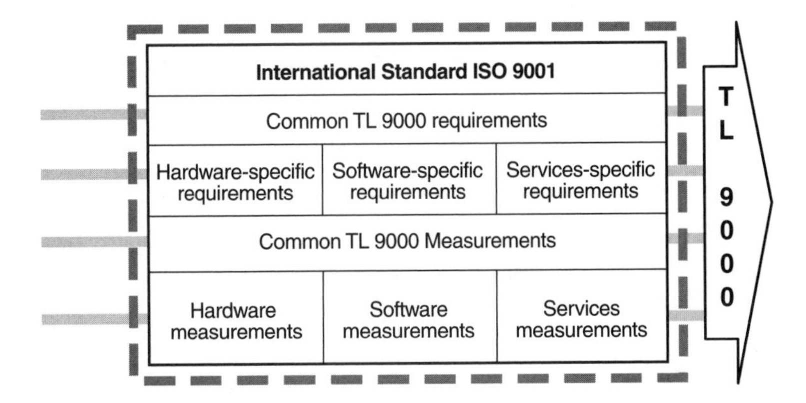

Figure 1.1 The TL 9000 Model.[6]
(Reprinted with permission of the QuEST Forum.)

SOURCES USED TO DEVELOP THE TL 9000 REQUIREMENTS HANDBOOK

At the start of TL 9000 development, there were many industry documents to use as sources. The following are the publications that were chosen as sources for the adders:

- ANSI/ISO/ASQC Q9001-2000,[1] the basic quality management standard. Sections 4–8 of this document contain requirements that were included verbatim in the *Requirements Handbook.*
- ANSI/ISO/ASQ Q9000-3,[3] the ISO document that provides guidance for applying ISO 9001 to software.
- ISO/IEC 12207,[4] the international standard that describes quality management system software lifecycle requirements.
- TR-NTW-000179,[5] the Telcordia Technologies document that describes telecommunications software requirements.
- GR-1202,[8] the Telcordia Technologies document that describes the Telcordia Customer Supplier Quality Program (CSQP℠).
- GR-1252,[9] the Telcordia Technologies document that describes telecommunications hardware requirements.
- ISO 9004-2,[10] the ISO document that provides quality management system guidelines for service organizations.

Table 1.1 The Sources of the Adders.

Source	Number of Requirements	Number of Notes
GR-1202	15	0
GR-1252	21*	0
TR-NWT-000179	13*	0
ISO/IEC 12207	13**	1
ISO 9000-3	6	1
ISO 9004-2	1	0
QuEST Forum Membership	14	0
Total	**81^**	**2**

Notes:
* This source contributed to one requirement jointly with one other source.
** This source contributed to two requirements jointly with one other source.
^ Two requirements were developed from multiple sources (two sources in each case).

In addition, the QuEST Forum membership provided an added set of requirements and QS-9000,[2] ISO 10011-1,[11] ISO 10011-2,[12] and ISO 10011-3[13] provided inputs for the appendixes.

Table 1.1 contains a summary of the number of contributions to the *Requirements Handbook*[6] from each source. GR-1252, the Telcordia Technologies hardware requirements document, provided the largest number of inputs. However, three software requirements documents, TR-NWT-000179, ISO/IEC 12207, and ISO 9000-3, collectively provided 32 requirements or parts of requirements.

KEY ADDITIONS TO ANSI/ISO/ASQC Q9001-1994

The following areas indicate where the QuEST Forum focused its attention and expanded on the current ISO 9001 requirements:

- Top management responsibilities
- Proper/robust planning:
  - Quality planning
  - Project planning
  - Configuration management planning
  - Product planning
  - Lifecycle planning
  - Test planning

- Customer–supplier communication requirements
- Quality improvement and customer satisfaction requirements
- Specialized service function requirements
- Certification measurements:
    - Objectives include a requirement to define measurement targets
    - Measurements must be defined, tracked, and reported to a central database
    - Measurements must be used to foster continual quality improvement
    - Auditors responsibilities with respect to measurements:
        - Assuring the suppliers' measurements processes are working
        - Assuring the information from the measurements are used by the suppliers to improve products, services, and processes
- Unlike ISO 9001, *should* means the preferred approach
    - Suppliers choosing other approaches must be able to demonstrate that their approach meets the intent of TL 9000
- Training sanctioned by the QuEST Forum and delivered by two authorized training providers. Currently the course providers are the Excel Partnership and STAT-A-MATRIX (The SAM Group).

Improvements in ANSI/ISO/ASQ Q9001-2000 Relating to Key TL 9000 Adders

The changes to ANSI/ISO/ASQ Q9001-2000 expanded requirements related to most of the TL 9000 additions. These changes were influenced by the QuEST Forum participation in the development of the revised standard as a liaison member of ISO/TC 176. The following are some examples of these changes[1]:

- More emphasis on top management responsibilities: "5.1 Management commitment. Top management shall provide evidence of its commitment to the development and implementation of the quality management system and continually improving its effectiveness by . . ."
- More focus on proper/robust planning: "7.1 Planning of product realization. The organization shall plan and develop the processes needed for product realization."

- Requirements for customer–supplier communication: "7.2.3 Customer communication. The organization shall determine and implement effective arrangements for communicating with customers."
- Requirements emphasizing quality improvement and customer satisfaction: "8.5.1 Continual improvement. The organization shall continually improve the effectiveness of the quality management system through the use of the quality policy, quality objectives, audit results, analysis of data, corrective and preventive actions and management review."
- Requirements covering specialized service functions: "7.2.1 Determination of requirements related to the product. The organization shall determine: a) requirements specified by the customer, including the requirements for delivery and *post-delivery activities*" (emphasis added). Also, "7.5.1 Control of production and service provision. The organization shall plan and carry out production and service provision under controlled conditions. Controlled conditions shall include, as applicable: f) the implementation of release, delivery and *post-delivery activities*" (emphasis added).
- Measurements required for certification: "5.4.1 Quality objectives. Top management shall ensure that quality objectives, including those needed to meet requirements for product [see 7.1 a)], are established at relevant functions and levels within the organization. The quality objectives shall be measurable and consistent with the quality policy." Also, "8.4 Analysis of data. The organization shall determine, collect and analyze appropriate data to demonstrate the suitability and effectiveness of the quality management system and to evaluate where continual improvement of the effectiveness of the quality management system can be made."

Although these changes strengthened ISO 9001 in areas of importance to the telecommunications industry, the working groups felt the need to include almost all of the adders from Release 2.5 in Release 3.0. The one change that best illustrates the shortfall is the requirements relating to specialized service functions. These functions usually happen for the industry in post-delivery activities. Although ANSI/ISO/ASQ Q9001-2000 includes references to post-delivery activities in the second half of two sentences in clauses 7.2.1 and 7.5.1, the effect on the related TL 9000 adders was negligible.

THE USE OF MEASUREMENTS TO IMPROVE PRODUCT AND SERVICE QUALITY

The unique difference between TL 9000 and current ISO 9001 practice is the use of measurements in continual quality improvement. These are measurements used to determine the quality level of products and services. Examples are circuit pack return rates, number of problem reports, and software update quality. It is expected that extending ISO 9000 in this direction will result in an industry drive to improve overall quality. Certainly, when suppliers know where their products stand relative to their competitors, and their customers know the quality level of "best in industry," then there will be a drive by all to improve.

The QuEST Forum envisions multiple uses for the measurements. For example, they can be used to enhance customer–supplier communication, prioritize and solve the most costly problems, and bridge the gap between quality issues and business results. Some service providers have indicated that they will use the measurements to develop report cards for their suppliers and some suppliers are using the measurements as part of their balanced scorecards.

The Forum has retained the University of Texas at Dallas (UTD) to gather the data and develop statistics for each product class. The statistics will identify the mean (or median) standard deviation (or range) and "best in industry." The data will be published on the QuEST Forum Web site (www.questforum.org). UTD was selected because of its long-standing role in telecommunications engineering education. It's expected that UTD will provide insight into benchmarking industry quality problems, identifying quality trends, and identifying possible solutions to quality problems.

The Measurements Implementation Process

The supplier gathers the data on a monthly basis and sends three months of data to UTD for inclusion in the industry database. A double blind system has been established to insure that the data is secure and not available to competitors or other unauthorized persons. The process works as follows:

- When a supplier has gathered the data, a notification is sent to the Forum Administrator, the American Society for Quality (ASQ), who provides encryption keys to the supplier to be used in sending the data to the industry database at UTD.

- The supplier encrypts the data and sends it to the Measurements Repository System (MRS) at UTD. Personnel transfer the data from the receiving server into the industry database. At least two individuals must be present during the transfer to ensure security of the data.

A major concern of the QuEST Forum membership has been the security of the data being supplied to UTD. The double blind process ensures that the data will be safe from access by competitors. In addition, no single individual at UTD will be able to access the data alone. The whole process has been developed with a minimum cost to suppliers.

UTD has complied with BS 7799,[14] the information security management requirement document developed by the British Standards Institute (BSI). UTD is the first U.S. organization to be certified to this document by BSI. Another part of the measurements process, the Registration Repository System (RRS) has also been certified to BS 7799.

Approved Measurements

Eleven measurements in five categories have been approved. These are as follows:[7]

1. Common to hardware, software, and services:
   - Number of problem reports (NPR)
   - Problem report fix response time (FRT)
   - Overdue problem report fix responsiveness (OFR)
   - On-time delivery (OTD)
2. Hardware and software only:
   - System outage measurement (SO)
3. Hardware only:
   - Return rates (RR)
4. Software only (software installation and maintenance):
   - Corrective patch quality (CPQ)
   - Feature patch quality (FPQ)
   - Software update quality (SWU)
   - Release application aborts (RAA)
5. Service only:
   - Service quality (SQ)

A major issue during the development of these measurements was the identification of normalization factors that allow comparison of similar products with different characteristics. For example, for circuit switching, the normalization factor for return rate is returns/10,000 terminations/year. Thus products with a large number of terminations per circuit pack are measured fairly against those with fewer terminations per circuit pack.

Another issue was a result of a requirement of some service providers that the suppliers provide Reliability and Quality Measurements for Telecommunications Systems (RQMS) measurements defined in the Telcordia Technologies document GR-929-CORE.[15] In some cases the supplier may provide the RQMS measurements in place of the corresponding TL 9000 measurements.

Other measurements and indicators are under consideration, but have not been approved. Indicators are a separate category of measures that are used to "flag potential cost, schedule, productivity, and quality issues," but are not reportable to UTD.[16]

The registrar has an important role in assuring compliance to the measurements requirements. "This includes assuring data validity and integrity, verifying [that] the benchmark measurements have been reported within the required time frame, and [assuring] that any measurements-specific nonconformance is resolved in a timely manner."[17] Also, because of the requirement in the *Requirements Handbook* that "objectives for quality shall include targets for the TL 9000 metrics defined in the *TL 9000 Quality System Measurements Handbook*,"[6] the registrar has a role in assuring this use of the measurements. Finally, there are requirements in the *Requirements Handbook* for the supplier to collect and analyze customer satisfaction (adder 8.2.1.C.1), field performance (adder 8.4.H.1), and service performance data (adder 8.4.V.1), some of which is measurements data.[6]

It should be noted that ANSI/ISO/ASQ Q9001-2000[1] includes Section 8.4 on "Analysis of Data."

The organization is required "to determine, collect and analyze appropriate data to demonstrate the suitability and effectiveness of the quality management system and to evaluate where continual improvement of the effectiveness of the quality management system can be made." This includes collecting data on customer satisfaction (Section 8.2.1) and conformity to customer requirements including delivery and post-delivery activities (Section 7.2.1). The registrar will assure that these requirements are satisfied in conjunction with the requirements to provide TL 9000 measurements.

PRODUCT AND SERVICE CATEGORIES

The initial set of product and service categories was provided in Appendix A of the *Measurements Handbook.*[7] Official product and service categories are subject to change. Because of the possibility of change, they are stored on the QuEST Forum Web site (www.questforum.org). The following are the product and service categories defined for the measurements implementation as of December 31, 2001:[7]

- *Switching equipment* broadly defined to include packet and circuit switched architectures
- *Signaling equipment* consists of five basic categories: supervisory, information, address, control, and alerting
- *Transmission equipment* connects the switched and interoffice networks with individual customers
- *Operations and maintenance equipment* used to manage, upkeep, diagnose and repair communications networks
- *Common systems*, including power systems and network equipment–building systems (NEBS), that provide space and environmental support for the network[18]
- *Customer premises equipment* commonly installed at the subscriber's location
- *Services,* which include installation, engineering, maintenance, repair, customer support, procurement, logistics, and general support services
- *Components and subassemblies* including products sold by component suppliers, contract manufacturers, and OEM suppliers

PILOT PROGRAM

One of the QuEST Forum initiatives was the creation of a pilot program to validate TL 9000 and its associated activities. The goals of the program were to provide feedback on the:

- Bottom-line value of the "adders"
- Processes that form the Measurements Reporting System
- Processes used to manage the registrar accreditation and supplier assessment/certification system

- Sanctioned training courses
- Audit ability of the requirements and measurements

Feedback was accomplished through weekly conference calls, minutes posted on the Web site (www.questforum.org), monthly QuEST Forum meetings, documented lessons learned, case studies, and commonly asked questions posted on the Web site.

Sixteen organizations from 11 companies participated in the program in four categories. Eight were hardware suppliers, three were software suppliers, four supplied hardware and software products, and one was a supplier of services. The participants are shown in Table 6.1, chapter 6.

The first step for each pilot organization was to make arrangements for its registrar to participate in the pilot program. Each registrar had to have its auditors take the QuEST Forum–sanctioned training and pass the auditor exam. Also, at least one member of the registrar's certification board with veto power or a majority of the board had to take the training and pass the exam. The registrar's accreditation agency also had to have at least one member of the registration decision-making body take the QuEST Forum–sanctioned training and pass the written exam. In addition, the accreditation agency had to witness at least one audit performed by each registrar using auditors who have completed the training and passed the exam.

Each pilot organization had to train its auditors and implementers, perform a gap analysis, develop processes for collection of the measurements, and submit the measurements data to UTD. The organizations that completed the process and were certified were not allowed to advertise the fact until the end of January 2000. They were honored at the January 25, 2000 Forum meeting in Dallas, Texas.

REGISTRATION PROCESS

The formal registration process started in the first quarter of 2000. The QuEST Forum established requirements for accreditation bodies to qualify registrars who will carry out TL 9000 certifications.[6] A supplier must demonstrate conformance to TL 9000 by successfully completing a third-party certification audit from an accredited TL 9000 registrar. The supplier will obtain a certificate in any combination of hardware, software, and/or services:

- Hardware: TL 9000-HW
- Software: TL 9000-SW
- Service: TL 9000-SC

Note that the supplier must be ISO 9001 certified or must include certification to ISO 9001 as part of the TL 9000 certification.

The QuEST Forum worked with the Registrar Accreditation Board (RAB) and the Standards Council of Canada (SCC) to pilot the procedures. The RAB and SCC oversee the competency of quality management systems certification bodies and environmental management systems certification bodies in their countries. Other accreditation agencies, such as the members of the International Accreditation Forum, will be invited to participate in the future. The following are the Web sites for these organizations:

- RAB (www.rabnet.com)
- SCC (www.scc.ca)
- IAF (www.iaf.nu)

TRAINING

The QuEST Forum has authorized two organizations, STAT-A-MATRIX (The SAM Group) and Excel Partnership, to develop and provide sanctioned TL 9000 training. The Forum retains the rights to the training intellectual property.

By selecting two training organizations to jointly provide training, the Forum expanded its philosophy of having competitors cooperate for the good of the industry. The two organizations should be commended for their joint effort and sharing of responsibilities and products. Also, the Forum should be commended for its foresight in taking this approach to providing sanctioned training. The results were better because two different perspectives were blended into the courses, making them stronger than the courses each trainer could have provided separately.

The two training providers have developed the following sanctioned courses:

1. *TL 9000 Quality Management System Overview:* A half-day to one-day course aimed at the majority of company associates, including senior management, that describes the goals of TL 9000 and how it can aid the organization.

2. *TL 9000 Quality Management System Implementation:* A three-day course that provides a road map for successful TL 9000 implementation and compliance.

3. *TL 9000 Quality Management System Auditing:* A four-day course that provides auditor training on the quality management system "adders" and the measurements requirements. A written exam is given at the completion of the course. Passage of the course and exam is required to be certified as a TL 9000 auditor.

4. *TL 9000 Quality Management System Auditing for Registrars:* A three-day course designed for auditors from registrar organizations. A test is given at the completion of the course. Passage is required to be certified as a TL 9000 registrar auditor.
5. *TL 9000 Quality Management System Measurements:* This course describes the details of submitting data to the database at UTD.
6. *ANSI/RAB/NAP Accredited Internal Auditing for the Telecommunications Industry:* This three-day course is accredited in the ANSI-RAB National Accreditation Program (NAP) and meets the training requirements for Quality Management Systems Internal Auditor certification. Passage of an exam is required.
7. *ANSI/RAB/NAP Accredited Lead Auditor Training for the Telecommunications Industry:* This three-day course is accredited in the ANSI-RAB National Accreditation Program (NAP) and meets the training requirements for Quality Management Systems Internal Lead Auditor certification. Passage of an exam is required.

The pilot organizations took early versions of the courses as part of the overall pilot program. They provided valuable input to the course developers that were later used to improve the courses.

ADMINISTRATION

An executive board consisting of six service providers and six suppliers governs the Forum. The chairmanship rotates on an annual basis between service providers and suppliers. The first chairman was Steve Welch, President, SBC Procurement/Corporate Real Estate. Krish Prabhu, President and Chief Operating Officer, Alcatel Telecom USA, succeeded him in 2000. The chairman in 2001 was George Via, Senior Vice President, Operations, Verizon; and the chairman in 2002 is Olga Striltschuk, Corporate VP and Director, Global Telecom Solutions Sector, Motorola.

The Forum selected the American Society for Quality (ASQ) to be the Forum Administrator and the University of Texas at Dallas (UTD) to administer the Measurements Repository System (MRS). One reason for the selection of ASQ was its historical role in support of quality standards development that includes providing the secretariat for the United States Technical Advisory Groups to ISO/TC 176, ISO/TC 207, ISO/TC 69 and IEC/TC 56. Also, ASQ's role as a worldwide leader in the quality profession played an important part in the selection.

ASQ is responsible for many functions, including the following[19]:

- Performance of general business functions
- Management of the program, including: QuEST Forum and working meetings; QuEST Forum board meetings and conference calls; membership and Web site administration
- Management of accreditation bodies and registrar information
- Administration of membership responsibilities
- Distribution and publishing of marketing materials
- Administration of training and education responsibilities
- Publication and distribution of the *Quality System Requirements* and *Measurements* handbooks and promotion of their use
- Arrangement of technical conferences and workshop programs

The Erik Jonsson School of Engineering and Computer Science at the University of Texas at Dallas was selected to administer the Measurements Repository System. The responsibilities include storage, security, maintenance, and statistical analysis of the data. It is also expected that the university will provide research into the solutions of industry problems identified as a result of the analysis.

Information Sources

The Forum's Web site (www.questforum.org) is the main source of information. The public section of the Web site contains information on membership, meetings, and ordering books, materials, and merchandise. The member section contains the measurements database summary results, individual membership information, presentations, minutes, product and service categories, and other QuEST Forum proprietary materials.

THE QUEST FORUM'S GLOBAL APPROACH

The Forum has made great strides in its goal to become a truly global telecommunications organization. It is the only global organization dealing with telecommunications quality management. Membership has grown internationally and includes service providers from the United Kingdom, South Africa, Australia, and Canada; suppliers from the United Kingdom, France, Israel, Canada, and many multinational suppliers based in the United States. In addition, the QuEST Forum has established liaison status with organizations in Europe, Russia, China, Taiwan, Japan, Korea, and Argentina.

The Forum has Liaison A status with ISO/TC 176, IEC/TC 56 and ISO/IEC JTC1/SC7. As a liaison member of ISO/TC 176, the Forum provided four members of task groups that worked on ANSI/ISO/ASQC Q9001-2000 and ANSI/ISO/ASQC Q9004-2000. These individuals contributed to the verification, validation, transition, and implementation processes used by ISO/TC 176.

The Forum has also developed a cooperative relationship with the EIRUS (European In-Process Quality Metrics [IPQM] and RQMS User Group) organization, a European telecommunication metrics user group. EIRUS uses two sets of metrics based on Telecordia Technologies requirements: European In-Process Quality Metrics (E-IPQM), and European Reliability and Quality Measurements for Telecommunications Systems (E-RQMS).[20] Joint working groups have been established to align the EIRUS and QuEST Forum measurements.

The QuEST Forum held meetings in Brussels in 1999; Paris, Yokohama, and London in 2000; and Buenos Aires and Berlin in 2001. Future meetings are scheduled for Seoul, Rio de Janeiro, and either Rome or Madrid in 2002.

As of December 31, 2001, there were 141 registered organizations. The United States led the way with 69, followed by the Asia/Pacific Region with 47, with the remainder in Canada, Europe, and Latin America.

As part of the global approach, the handbooks have been translated into Chinese, and the TL 9000 Overview, TL 9000 Implementation, and TL 9000 Auditing courses have been translated into Korean, Portuguese (Brazil), and Spanish (Mexico). Translations into other languages are under consideration.

ENDNOTES

1. ANSI/ISO/ASQ Q9001-2000, *Quality Management Standards—Requirements*, 3rd ed. (Milwaukee: ASQ Quality Press, 2000).
2. Automotive Industry Action Group, *Quality System Requirements, QS-9000*, 3rd ed. (Detroit, MI: Automotive Industry Action Group [AIAG], 1998).
3. ANSI/ISO/ASQ Q9000-3-1997, *Quality Management and Quality Assurance Standards—Part 3: Guidelines for the Application of ISO 9001:1994 to the Development, Supply, Installation, and Maintenance of Computer Software*, 2nd ed. (Milwaukee: ASQ Quality Press, 1997).
4. ISO/IEC 12207, *Information Technology Software Life Cycle Processes* (Geneva, Switzerland: International Organization for Standardization, 1995).

5. TR-NWT-179, *Quality System Generic Requirements for Software,* no. 2 (Morristown, NJ: Telcordia Technologies, 1993).
6. The QuEST Forum, *TL 9000 Quality Management System Requirements Handbook*, Release 3.0 (Milwaukee: ASQ Quality Press, 2001).
7. The QuEST Forum, *TL 9000 Quality Management System Measurements Handbook*, Release 3.0 (Milwaukee: ASQ Quality Press, 2001).
8. GR-1202-CORE, *Generic Requirements for Customer Sensitive Quality Infrastructure,* no. 1 (Morristown, NJ: Telcordia Technologies, 1995).
9. GR-1252, *Quality System Generic Requirements for Hardware,* no. 1 (Morristown, NJ: Telcordia Technologies, 1995).
10. ANSI/ISO/ASQC Q9004-2-1994, *Quality Management and Quality System Elements—Part 2: Guidelines for Services* (Milwaukee: ASQC Quality Press, 1994).
11. ANSI/ISO/ASQC Q10011-1-1994, *Guidelines for Auditing Quality Systems—Auditing* (Milwaukee: ASQC Quality Press, 1994).
12. ANSI/ISO/ASQC Q10011-2-1994, *Guidelines for Auditing Quality Systems—Qualification Criteria for Quality Systems Auditors* (Milwaukee: ASQC Quality Press, 1994).
13. ANSI/ISO/ASQC Q10011-3-1994, *Guidelines for Auditing Quality Systems—Management of Audit Programs* (Milwaukee: ASQC Quality Press, 1994).
14. BS 7799:1999, *Information Security Management, Part 1: Code of Practice for Information Security Management and Part 2: Specification for Information Security Management* (London: The British Standards Institute, 1999).
15. GR-929-CORE, *Reliability and Quality Measurements for Telecommunications Systems (RQMS),* no. 4 (Morristown, NJ: Telcordia Technologies, 1998).
16. Henry Malec, "TL 9000 Database Repository and Metrics," *The Informed Outlook* 4, no. 6 (June 1999): 4.
17. Galen Aycock, Jean-Normand Drouin, and Thomas Yohe, "TL 9000 Performance Metrics to Drive Improvement," ASQ *Quality Progress* 32, no. 7 (July 1999): 41–45.
18. GR-63-CORE, *Network Equipment-Building System (NEBS) Requirements: Physical Protection,* no. 1 (Morristown, NJ: Telcordia Technologies, 1995).
19. "Development of ASQ Role as TL 9000 Forum Administrator," *The Informed Outlook* 4, no. 5 (May 1999): 1–2.
20. Information concerning the European Quality Metrics: E-IPQM and E-RQMS, can be obtained from EURSECOM GmbH, Scloss Wolfsbrunnenweg 35, 69118 Heidelberg, Germany, and its Web site, www.eurescom.de/.

2

Common Requirements

INTRODUCTION

The TL 9000 requirements are based on ANSI/ISO/ASQ Q9001-2000[1] and 81 additional requirements or "adders," which are included to either enhance the quality system or be specific to the telecommunications industry. It should be noted that some of the requirements of ANSI/ISO/ASQ Q9001-2000 were influenced by inputs from the QuEST Forum representatives to ISO/TC 176, resulting in a closer tie between TL 9000 and ANSI/ISO/ASQ Q9001-2000.

The adders are described in this chapter and chapter 3.* In this chapter we discuss the requirements that are common to hardware, software, and service organizations. Chapter 3 contains a discussion of hardware only, software only, service only, and paired combinations of hardware, software, and service requirements. In both chapters we repeat the requirements from the *Requirements Handbook,*[2] give a rationale for each requirement, and provide "tips" from organizations that have implemented TL 9000. The reader should focus on the tips because they provide real information on how some organizations implemented the requirements.

The standard definition for the word *shall* in ANSI/ISO/ASQ Q9001-2000 is used in TL 9000, but *should* has a different meaning. The word *shall* is used for mandatory requirements that must be adhered to; *should,*

---

*The adders and notes in chapters 2 and 3 are repeated verbatim with permission of the American Society for Quality, publisher of the *TL 9000 Quality Management System Requirements Handbook* for the QuEST Forum.

on the other hand, indicates "the preferred approach and organizations choosing other approaches must be able to show that their approach meets the intent of TL 9000."[2]

CODES USED TO IDENTIFY THE TYPE OF REQUIREMENT

Individual codes were given to identify each adder as hardware-, software-, or service-specific, or when the adder is common to hardware, service, or software:

C: Common (hardware, software, and service)

HS: Hardware and software

HV: Hardware and service

H: Hardware only

S: Software only

V: Service only

COMMON ADDERS

Common adders affect the following elements. The number of adders in each element is shown in parenthesis. There are a total of 39 common adders:

- *4 Quality management system:* 4.2 Documentation requirements (1)
- *5 Management responsibility:* 5.2 Customer focus (2); 5.4 Planning (4); 5.5 Responsibility, authority and communication (1)
- *6 Resource management:* 6.2 Human resources (6); 6.4 Work environment (1)
- *7 Product realization:* 7.1 Planning of product realization (4); 7.2 Customer-related process (4); 7.3 Design and development (7); 7.4 Purchasing (1); 7.5 Production and service provision (3)
- *8 Measurement, analysis and improvement:* 8.2 Monitoring and measurement (2); 8.4 Analysis of data (1); 8.5 Improvement (2)

4.2.3 Control of documents

4.2.3.C.1
Control of
customer-supplied
documents and data

Adder: 4.2.3.C.1 Control of Customer-Supplied Documents and Data—The organization shall establish and maintain a documented procedure(s) to control all customer-supplied documents and data (e.g., network architecture, topology, capacity, installation termination assignments, and database) if these document and data influence the design, verification, validation, inspection and testing, or servicing the product.

Rationale: The customer may need to provide documents and data to assist in the development of the product. The documents and data may contain "proprietary" information and failure to institute proper controls of identification can result in legal problems and loss of customer confidence.

TIPS: The same procedures used for control of internal documents and data may be used for this requirement.

5.2 Customer focus

5.2.C.1 Customer relationship development	5.2.C.2 Customer communication procedures

Adder: 5.2.C.1 Customer Relationship Development—Top management shall demonstrate active involvement in establishing and maintaining mutually-beneficial relationships between the organization and its customers.

Rationale: Customer satisfaction and meeting customer needs are the key to the success of any organization. It is essential to have a robust relationship between the customer and top management in order to ensure that the customer's voice is heard in the development of products and the quality required. Many active user groups feed future product enhancement ideas to the design and development organizations. This feedback would be extremely advantageous in satisfying customer expectations.

TIPS: This adder requires that the organization's executives communicate directly with the customer. It can't be delegated. This may include customer visits, attendance at user-group meetings, or being active in the QuEST Forum.

Adder: 5.2.C.2 Customer Communication Procedures—The organization shall establish and maintain a documented procedure(s) for communicating with selected customers. The documented procedure(s) shall include:

a) a strategy and criteria for customer selection,

b) a method for the organization and its customers to share joint expectations and improve the quality of products, and

c) a joint review with the customer at defined intervals covering the status of shared expectations and including a method to track the resolution of issues.

Rationale: The telecom industry has a history of strong customer–supplier communication. This is because of the complexity of the products and the need to continually add features. This clause addresses the procedures for obtaining input from the customer. Regular customer communication ensures that the customer's needs are heard.

TIPS: Many companies have documented procedures for determining the customer's quality expectations. The procedures may include joint reviews. The feedback from the reviews is valuable in influencing new products or services, adding new features, and improving overall quality of the products.

Joint reviews at regular intervals will assist in uncovering defects or issues on an ongoing basis. This will assist in project management and will prevent "surprises" from showing up at the end.

The procedure in CSQP℠ is to hold semiannual meetings with key customers. The agenda includes strategies and change in product line.

5.2.C-NOTE 1:* It is recognized that it is not possible for an organization to provide the same level of communication with all its customers. The level provided may depend on the amount of business with the customer, the history of problems, customer expectations, and other factors

*Notes are suggestions only, not requirements.

(see Appendix F of *TL 9000*, Release 3.0, "Guidance for Communication with Customers").

This note provides guidance on how to choose methods of communication for different types of customers. Appendix F is an excellent source of information on the methods of communication.

5.4 Planning

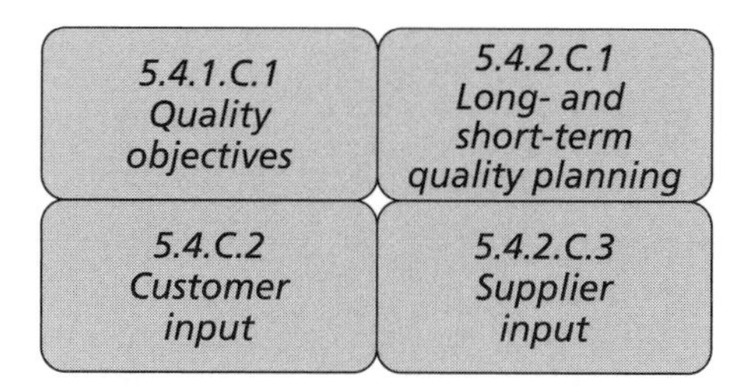

Adder: 5.4.1.C.1 Quality Objectives—Objectives for quality shall include targets for the TL 9000 measurements defined in the *TL 9000 Quality Management System Measurements Handbook.*[3]

Rationale: "If you don't know where you are going, how do you know you have arrived?"[4] In order to implement any process improvement, you need to know where you are today. For example, if you start a diet to lose 20 pounds, you need to know your current weight—the baseline—before you begin the diet. The purpose of this baseline number is to learn whether or not the diet is working for you. The same principle applies to the quality objectives. If quality objectives are not established, how can you tell that you have met your objectives? The metrics will indicate to the management whether the product or service quality is improving. This requirement will enable management to evaluate current processes, set goals, and create the means for improvement opportunities.

TIPS: Those metrics that relate to the product being certified must be defined in the organization's quality objectives, and there must be plans in place for the collection, validation, and reporting of the data. One organization had a five-year plan with objectives for TL 9000 metrics that link to the business plan.

Adder: 5.4.2.C.1 Long- and Short-Term Quality Planning—The organization's quality planning activities shall include long- and short-term plans with goals for improving quality and customer satisfaction. Performance to these goals shall be monitored and reported. These plans shall address:

a) cycle time,

b) customer service,

c) training,

d) cost,

e) delivery commitments, and

f) product reliability.

5.4.2.C.1-NOTE 1:* Top Management should demonstrate their active involvement in long- and short-term quality planning.

Rationale: The planning enables the organization to conduct risk assessments and identify mitigation activities, allocate appropriate resources, manage costs, and identify dependencies. Planning also forces the organization to evaluate training needs, delivery commitments, process improvements, and potential corrective actions. Project planning helps organizations ensure that schedules are met and target completion dates do not slip. Long-term planning addresses maintenance activities, customer support, and installation.

TIPS: NOTE 1 states that "Top management should demonstrate active involvement in long- and short-term quality planning." Executives should not continually delegate their authority, but should show that they periodically attend quality-planning sessions. A lifecycle model should be defined and enforced for product development and should encourage planning in every major phase. Tools such as plan–do–check–act (PDCA) should be used. Employee goals and objectives should cover items (b) through (e) and be tracked with employee reviews and appraisals. The organization's long- and short-term objectives must address each of the items and the use of customer satisfaction data as an input. Quality improvements could be in the form of reduced cycle time, quicker response time, availability of training, reduced costs, on-time delivery, and/or extended product life.

One company uses a two-level approach. Executives develop high-level critical objectives. The operational level translates these into implementation activities. They also use kaizen-type inputs from employees.

Adder: 5.4.2.C.2 Customer Input—The organization shall implement methods for soliciting and considering customer input for quality planning activities. The organization should establish joint quality improvement programs with customers.

---

*Notes are suggestions only, not requirements.

Rationale: Through the use of the products, the customer becomes familiar with the problems he/she has encountered and will have valuable feedback on the following aspects of the product: usability, maintainability, and performance. Regular customer involvement ensures that customer's requirements are fulfilled related to product design, features, and domestic or international standards. The customer involvement may be in the selection of subcontractors, conducting joint reviews, or identifying requirements.

TIPS: A customer satisfaction survey consisting of questions directly related to the usability, maintainability, and robustness of the products will give the organization the information required to implement quality improvement programs. Where feasible, setting quality goals and implementing improvement programs based on joint meetings with customers would be beneficial for the organization.

Hold periodic (at least twice yearly) meetings with individual customer(s) to share expectations, identify gaps, develop action plans, measure improvements, and review/update the action register. Meet at the customer's site where new product has been introduced within the first year to discuss product-specific issues.

An alternate method of communication is to hold periodic user-group meetings. This adder, along with 5.2.C.2 (Customer Communication Procedures) and 8.2.1.C.1 (Customer Satisfaction Data), caused the marketing department in one company to formally define and document a customer relationship process.

Adder: 5.4.2.C.3 Supplier Input—The organization shall implement methods for soliciting and using supplier input for quality planning activities.

Rationale: Telecom organizations are partnering more and more with contract manufacturers and other suppliers. When an organization purchases a product or service from a supplier, the organization needs to clearly specify requirements and establish a procedure to collect information on quality-related activities. Supplier feedback is related to practices, procedures, and activities that include product capabilities, design constraints, product validation requirements, or lessons learned. Communication between an organization and a supplier should clearly establish the capability of the supplier to meet the organization's requirements.

TIPS: In order to track the feedback, a formal method of collecting and documenting the feedback should be created by the organization.

During the development phase, a project plan should be written with inputs from suppliers to include resources and commitments. Also, the suppliers should provide technical exchange of information and ideas. This is a partnering effort. Where suppliers impact any of the quality planning items, there must be communications established with them.

5.5 Responsibility, authority and communication

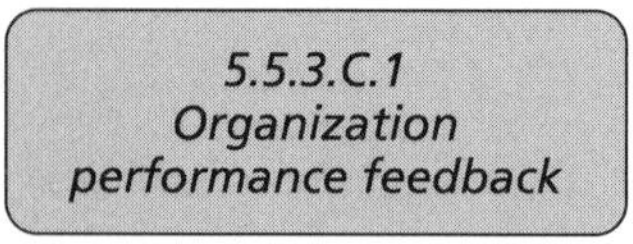

5.5.3.C.1
Organization performance feedback

Adder: 5.5.3.C.1 Organization Performance Feedback—The organization shall inform employees of its quality performance and the level of customer satisfaction.

Rationale: When employees know about customer satisfaction and quality performance results they feel a part of the business. Since their work has a direct relationship to quality performance, raising their awareness and participation will result in overall improvements.

TIPS: In many companies the customer satisfaction numbers are based on a rigorous statistical analysis, and the satisfaction number is tied to employee bonuses and other compensation. In these companies, trend analysis of customer satisfaction is kept and each year the "bar" for the satisfaction is raised. If the company reaches the preestablished satisfaction number, the staff is rewarded with bonuses or other compensation. Many companies use continuous customer satisfaction surveys in lieu of a yearly survey. One company displays customer report cards based on CSQP[SM5,6] on its Web site. They are also displayed throughout the organization. In some countries, surveys are not effective. In those countries, customer satisfaction data is obtained through customer meetings.

6.2 Human resources

6.2.2.C.1 Internal course development	*6.2.2.C.2 Quality improvement concepts*	*6.2.2.C.3 Training requirements and awareness*
6.2.2.C.4 ESD training	*6.2.2.C.5 Advanced quality training*	*6.2.2.C.6 Training content*

Adder: 6.2.2.C.1 Internal Course Development—When the organization develops internal training courses, it shall establish and maintain a process for planning, developing, and implementing these courses.

Rationale: For any type of training, planning is required. Some subjects to be considered in the training plans include:

- Objectives of the training
- Contents of the training
- Experience of audience in the subject matter
- Length of the training
- Schedule
- Training equipment
- Training facilities
- Piloted training and incorporation of the feedback to continually improve the contents of the training
- Establishing a formal process to keep a record of attendees who successfully complete the training
- Determining a method of reviewing the effectiveness of the training

If in-house-developed training has no structured process, it will become obsolete in a short period of time.

TIPS: Training departments should have a procedure on developing courses and maintaining records. The procedures should include a means of reviewing the effectiveness of the training. This may consist of on-the-job testing, course surveys of students, interviewing the students' managers, and review of process and product defects. A training curriculum can help establish "centers of excellence" or an expert network.

Adder: 6.2.2.C.2 Quality Improvement Concepts—Those employees that have a direct impact on the quality of the product, including top management, shall be trained in the fundamental concepts of continual improvement, problem solving, and customer satisfaction.

Rationale: Most employees in an organization have a direct impact on the quality of the product. It is especially important for management with executive responsibility to have quality awareness and understand the concepts of total quality management that will result in continual improvement of the processes and products. Understanding of quality concepts by all employees will affect the robustness of the products, and ultimately customer satisfaction.

TIPS: This adder requires that quality improvement concepts be part of most employee's training curriculum, including executives.

In one organization, a new employee has to have an overview of quality concepts, and every quarter a schedule on available courses goes to each employee. The overview course is given to supervisors, engineers, and managers. Supervisors cascade the training to the shop personnel. Senior leadership demonstrates commitment by attending class with their staff.

Adder: 6.2.2.C.3 Training Requirements and Awareness—Training requirements shall be defined for all positions that have a direct impact on the quality of products. Employees shall be made aware of training opportunities.

Rationale: Behind the success of every company is the experience and knowledge of the individuals who are part of the company. Depending on the job to be performed, each position within a company requires special skills. Identification of the skill set needed to accomplish a particular job and providing adequate training assists the employees in fulfilling their job-related responsibilities. Well-trained employees are the most valuable asset for the success of an organization.

TIPS: This adder requires that there be training requirements specified for each job position, and that these requirements be reviewed with employees.

Adder: 6.2.2.C.4 ESD Training—All employees with functions that involve any handling, storage, packaging, preservation, or delivery of ESD-sensitive products shall receive training in electrostatic discharge (ESD) protection prior to performing their jobs.

Rationale: ESD damages components and products such as integrated circuits, printed wiring board assemblies, magnetic tapes and/or disks, and other media used for software or data storage. In order to prevent this damage, employees should be provided with training so they will be aware of the ramifications of ESD and take precautionary measures to avoid damage.

TIPS: This adder requires ESD training prior to working on ESD-sensitive items.

Adder: 6.2.2.C.5 Advanced Quality Training—The organization shall offer appropriate levels of advanced quality training. Examples of advanced quality training may include statistical techniques, process capability, statistical sampling, data collection and analysis, problem identification, problem analysis, and corrective and preventive action.

Rationale: Every employee should be familiar with Deming's "plan–do–check–act" methodology:

- Planning for a new process
- Implementing the new process
- Collecting information and a statistical sampling of data to make sure the new process works
- Analyzing and planning a new approach if the sampling of data reflects no improvement

Product quality is affected by the skills and knowledge of the personnel.

TIPS: Training in process improvements, verification, and validation activities will have a direct impact on the quality of the products. Root cause analysis is a vital activity in identifying the causes of the problems, and the ability of the employees to address these causes will eliminate similar problems in the future.

This adder requires that training for advanced quality techniques be available, as appropriate. This does not mean the organization must develop these courses internally, but they must have a way of training employees in these techniques. Commercial training courses may be used.

Adder: 6.2.2.C.6 Training Content—Where the potential for hazardous conditions exists, training content should include the following:

a) task execution,

b) personal safety,

c) awareness of hazardous environment, and

d) equipment protection.

Rationale: Training should be given in order to eliminate and reduce employee injuries or damage to the business and related equipment. The training should specifically address:

- Personal safety to avoid injuries
- The process used to execute the job
- Awareness of hazardous environments or materials that might damage the product, equipment, and/or personnel

TIPS: This is a "should" requirement and may be replaced by an equivalent process that meets the intent of the requirement. This adder requires that there be training courses available for personal safety, hazardous material handling, and service protection. Every new employee should go through training on hazardous materials.

6.4 Work environment

6.4.C.1
Work areas

Adder: 6.4.C.1 Work Areas—Areas used for handling, storage, and packaging of products shall be clean, safe, and organized to ensure that they do not adversely affect quality or personnel performance.

Rationale: The working environment contributes to the overall quality of the products, as well as affecting the morale and well-being of the personnel. Poor working conditions may have an adverse impact on the product. External factors such as handling, storage, and packaging of products can contribute to the nonconformity of the product. Nonconforming returned products must be organized and stored adequately to eliminate erroneously returning them to a customer.

TIPS: This adder enforces the need for clean, safe, and organized work areas. It may be necessary to train personnel involved with development and manufacturing. Where "clean rooms" are involved, special environments must be maintained and special procedures followed.

7.1 Planning of product realization

7.1.C.1 Life cycle model	*7.1.C.2 New product introduction*	*7.1.C.3 Disaster recovery*	*7.1.C.4 End of life planning*

Adder: 7.1.C.1 Life Cycle Model—The organization shall establish and maintain an integrated set of guidelines that covers the life cycle of its products. This framework shall contain, as appropriate, the processes, activities, and tasks involved in the concept, definition, development, production, operation, maintenance, and (if required) disposal of products, spanning the life of the products.

Rationale: In today's environment, products and services are becoming complex. In multifaceted products the interdependencies and designs need to be considered up front to increase their serviceability, reliability, and availability. In order to ensure completeness of all activities and to make the product coherent, up-front identification of business-related risks, project dependencies, costs, and resources will prevent major surprises later on. This adder ensures that the organization considers all activities during the life of its products and how these activities will affect the overall quality.

TIPS: Utilization of a gate process* will satisfy this requirement, but it must include all elements of the entire lifecycle such as operation, support, and product disposal (when needed). It can't stop when the product is deployed and becomes generally available. Ongoing planning and support must be demonstrated. Also, it is critical that when a product or service is discontinued, there be a planning and notification process that includes all customers.

If your scope only covers manufacturing, you can define your lifecycle to cover manufacturing only. If parts of the lifecycle are outsourced (for example, contract manufacturing), the lifecycle must be linked into ANSI/ISO/ASQ Q9001-2000 elements 4.1 (General requirements) and 7.4 (Purchasing). Gaining a companywide agreement on the interpretation and action plan involves more than expected time and resources. Do not underestimate the impact of this element. Involve Engineering and Sales early on with this task.

Adder: 7.1.C.2 New Product Introduction—The organization shall establish and maintain a documented procedure(s) for introducing new products for General Availability.

7.1.C.2-NOTE 1:** The new product introduction program should include provisions for such programs as: quality and reliability prediction studies, pilot production, demand and capacity studies, sales and service personnel training, and new product post-introduction evaluations.

Rationale: The purpose of this procedure is to define the common process for introducing a new product. The process, when implemented, will ensure that when any new product is introduced, there is a consistent

*Gate process: a process in which at each stage, certain criteria must be met before the product can move to the next stage.

**Notes are suggestions only, not requirements.

process being followed. This will eliminate a need for each team to develop its own customized process.

TIPS: A checklist or a template may be developed with all necessary components including reliability, serviceability, and availability. The logistics may include customer support, sales, manufacturing, and service readiness.

Adder: 7.1.C.3: Disaster Recovery—The organization shall establish and maintain methods for disaster recovery to ensure the organization's ability to recreate and service the product throughout its lifecycle.

Rationale: Remember the San Francisco earthquake in October 1989 and the fire in a Con Edison substation at the foot of the Brooklyn Bridge. Communications were completely out for many hours after the earthquake and approximately 1000 companies were affected by the fire. A study by Contingency Planning Associates indicated that only 25 of the affected companies had any type of disaster recovery plan.

Disaster recovery is the most overlooked activity. With our growing reliance on the expanding technology, there certainly are increased risks when a disaster strikes. It is important for an organization to address disaster recovery planning so that the product and service can be supported without major interruptions. Certainly after the events of September 11, 2001 all companies should develop a disaster recovery plan.

TIPS: This adder requires that a recovery contingency plan be in place in the event that a disaster affects the organization's operations. A documented procedure(s) should be written to provide backup for software and storage off site of the tapes or disks. Also, there is a need to provide backup and support for key personnel whose loss would constitute a disaster. Natural disasters such as floods, tornadoes, and hurricanes have the ability to completely demolish a facility and these disasters should not be overlooked in terms of disaster recovery planning.

Adder: 7.1.C.4 End of Life Planning—The organization shall establish and maintain a documented procedure(s) for the discontinuance of manufacturing and/or support of a product by the operation and service organizations. The documented procedure(s) should include:

a) cessation of full or partial support after a certain period of time,

b) archiving product documentation and software,

c) responsibility for any future residual support issues,

d) transition to the new product, if applicable, and

e) accessibility of archive copies of data.

Rationale: It is critical for customers to know when an organization is going to stop supporting a particular product so they can decide to either buy a replacement product or plan in advance as to how the product will be maintained. By planning and letting customers know in advance when a product will be discontinued, the organization manages customer expectations.

TIPS: Remember, this is a "should" requirement and may be replaced by an equivalent process that meets the intent of the requirement. Item (a) requires that the organization notify their customers when a product's support is going to change. Items (b) and (e) require that the organization retain discontinued product documentation and software and that it be accessible. Item (c) requires that the organization plan for support of residual problems. Item (d) requires that the organization inform the customers about transition to new product.

In one company, the Business Management department handles end-of-life planning.

7.2.2 Review of requirements related to the product

There are no adders in this subsection. However, there are two notes that are intended to make this process more consistent across the industry.

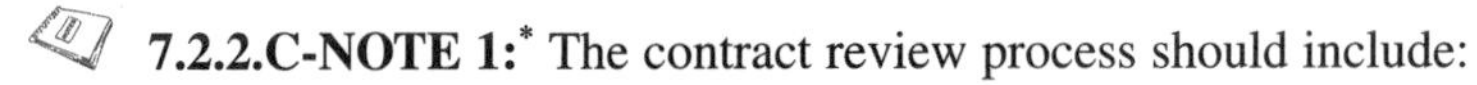

7.2.2.C-NOTE 1:* The contract review process should include:

a. Product acceptance planning and review;

b. Handling of problems detected after product acceptance, including customer complaints and claims; and

c. Responsibility of removal and/or correction of nonconformities after applicable warranty period or during product maintenance contract period.

TIPS: Item (a) is covered in detail by 7.2.2.C-Note 2. Item (b) requires details on how the customer notifies the organization of a problem. Item (c) deals with how problems are handled after the product warranty period has elapsed.

7.2.2.C-NOTE 2:* The product acceptance plan should include, as appropriate:

a. Acceptance review process;

b. Acceptance criteria;

*Notes are suggestions only, not requirements.

c. Documented test procedures;

d. Test environment;

e. Test cases;

f. Test data;

g. Resources involved;

h. Method(s) for problem tracking and resolution; and

i. Required acceptance test reports.

TIPS: Items (a), (b), and (c) require that documentation be provided with these details. Item (d) requires any special environmental conditions, such as temperature control, heat tests, and so on, be documented. Items (e) and (f) require that data from previous tests be made available. Item (g) requires providing resource needs for acceptance tests. Item (h) requires showing how problems are resolved and tracked, using tools such as mechanized diagnostics. Item (i) requires showing a sample of what an acceptance report will look like.

7.2.3 Customer communication

7.2.3.C.1 Notification about problems	*7.2.3.C.2 Problem severity*	*7.2.3.C.3 Problem escalation*	*7.2.3.C.4 Customer feedback*

Adder: 7.2.3.C.1 Notification About Problems—The organization shall establish and maintain a documented procedure(s) to notify all customers who may be affected by a reported problem that is service affecting.

Rationale: Early notification to customers of any service-affecting problems will assist in reduced customer calls and will keep the customers abreast of quality-related issues with the product. The proactiveness on the organization's part builds customer confidence.

TIPS: This adder requires that there be a customer notification process for reporting service-affecting problems. Document a procedure on how to contact the customers when a problem occurs. This will assist the organization in managing customer anxiety and frustration.

Adder: 7.2.3.C.2 Problem Severity—The organization shall assign severity levels to customer-reported problems based on the impact to the customer in accordance with the definitions of critical, major, and minor problem reports contained in the glossary of this handbook. The severity level shall be used in determining the timeliness of the organization's response.

7.2.3.C.2-NOTE 1:* The customer and the organization should jointly determine the priority for resolving customer-reported problems.

Rationale: A problem that may be considered critical by a customer may not necessarily be perceived with a similar criticality by the organization. Having the organization and customer jointly identify and understand the criticality of the problem will assist in eliminating the waiting time frustration of the customer. Understanding how a defect would affect the customer's ability to provide service will allow the organization to allocate adequate resources to resolve a defect in an efficient manner.

TIPS: This adder requires that there be customer input in determining the severity of a reported problem and that the severity be based upon customer impact. At one company's Technical Assistance Center, employees receive customer complaints and assign severity per customer suggestions. At times, subject matter experts review the severity assignment and negotiate a change with the customer.

Adder: 7.2.3.C.3 Problem Escalation—The organization shall establish and maintain a documented escalation procedure(s) to resolve customer-reported problems.

Rationale: When a problem cannot be resolved by the first-level customer-support personnel, it is important for the organization to have a documented procedure on escalation to a second-level support organization so it can be resolved in a timely fashion. In many companies, there are multiple levels of support organizations and a problem may be transferred from one level to the next when it cannot be resolved by the initial level of support.

TIPS: Have the appropriate support people on call. In one organization, problems are documented at the customer assistance center where they try to resolve them online. If they can't solve them, the problems are escalated to the appropriate organizations.

---

*Notes are suggestions only, not requirements.

Adder: 7.2.3.C.4 Customer Feedback—The organization shall establish and maintain a documented procedure(s) to provide the customer with feedback on their problem reports in a timely manner.

Rationale: Timely resolution of customer-reported problems is important for customer satisfaction. Even if the problem takes longer than expected to resolve, keeping the customer abreast of its status will increase customer confidence. A documented procedure for giving the customer feedback on reported problems will enable the organization to have a history of the problem.

TIPS: An automated defect tracking system with status and comments on each problem will provide an analysis of resolution times and historic information.

7.3 Design and development

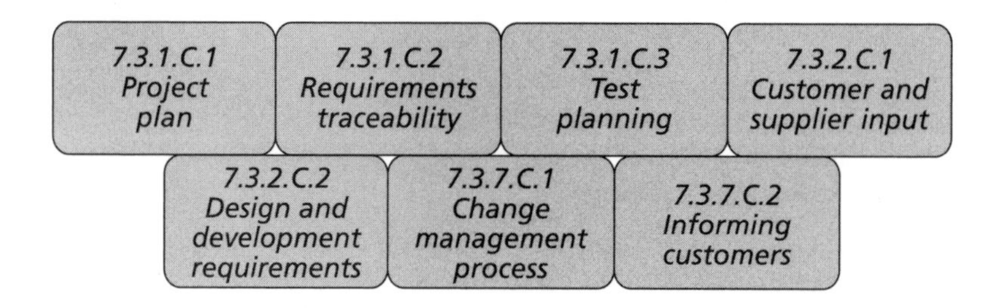

Adder: 7.3.1.C.1 Project Plan—The organization shall establish and maintain a project plan based on the defined product life cycle model. The plan should include:

a) project organizational structure,

b) project roles and responsibilities,

c) interfaces with internal and external organizations,

d) means for scheduling, tracking, issue resolution, and reporting,

e) budgets, staffing, and schedules associated with project activities,

f) method(s), standards, documented procedure(s), and tools to be used,

g) references to related plans (e.g., development, testing, configuration management, and quality),

h) project-specific environment and physical resource considerations (e.g., development, user documentation, testing, and operation),

i) customer, user, and supplier involvement during the product life cycle (e.g., joint reviews, informal meetings, and approvals),

j) management of project quality,

k) risk management and contingency plans (e.g., technical, cost and schedules),

l) performance, safety, security, and other critical requirements,

m) project-specific training requirements,

n) required certifications,

o) proprietary, usage, ownership, warranty, licensing rights, and

p) post-project analysis.

7.3.1.C.1-NOTE 1:* The project plan and any related plans may be an independent document, a part of another document, or comprised of several documents.

7.3.1.C.1-NOTE 2:* General work instructions defining tasks and responsibilities common to all development projects need not be replicated as part of a project plan.

Rationale: A project plan identifies major milestones of the entire project, from beginning to end, and also identifies individual work breakdown activities that must be completed to reach these milestones. It's important to control slippage in the schedule and identify and resolve issues before they impact the project adversely. The project plan defines resources and schedules, identifies risks, indicates activities assigned to appropriate resources, and highlights dependencies. Individual functional groups within a project may have their own mini–project plan that links into an overall master project plan.

Development plan, test plan, configuration management plan, support plan, risk identification/mitigation plan, and security plan may be part of the project plan or separate documents within the project.

---

*Notes are suggestions only, not requirements.

TIPS: Remember, this is a "should" requirement and may be replaced by an equivalent process that meets the intent of the requirement. Use two levels in project plans. The highest level covers all current activities. The second level covers major milestones, work breakdown activities, and schedules. For each activity, a template can be used to cover the key points.

Items (a) through (e) require the organization to show that there are adequate human resources planned for the project. Items (f), (g), and (h) require the organization to show that there are adequate documents and tools planned to support the project. Item (i) requires the organization to show that there has been involvement outside the design group, such as customer, user, and supplier inputs. Item (j) requires the organization to show the project quality plan. Item (k) requires the organization to show what contingency will be used if the project resources require change. Item (l) requires that the organization address these specific issues within their requirements. Item (m) requires that the organization provide training for the project. Item (n) requires the organization to identify any special certifications needed. Item (o) requires the organization to identify all legal aspects of the project. Item (p) requires the organization to review the results of the finished project and analyze the results. In the project plan, identify each major development area, assign personnel, and plan deliverables, schedules, and resources. Track resources and schedules during development. Use the gate lifecycle model to establish project controls.

The notes attached to this adder provide the organization with the latitude to reference the same document(s) for several different projects.

Adder: 7.3.1.C.2 Requirements Traceability—The organization shall establish and maintain a method to trace documented requirements through design and test.

7.3.1.C.2-NOTE 1:* The organization should establish communication methods for dissemination of product requirements and changes to requirements to all impacted parties identified in the project plan.

Rationale: Often when a product or service progresses from one phase to the next, some of the initial requirements are lost along the way. Traceability is often needed because of legal or statutory requirements, hazardous materials, or individual status of component parts. Effective traceability confirms that all requirements are present in the finished product and can be accounted for from initial phase through design and development to the finished product.

*Notes are suggestions only, not requirements.

TIPS: During the design and implementation process, requirements and design documents show mapping of design requirements through subsequent phases. An automated tool can be used for creating the mapping. Design reviews may be performed at various stages to verify that the design process is not deviating from the specified requirements.

Adder: 7.3.1.C.3 Test Planning—Test plans shall be documented and results recorded. Test plans should include:

a) scope of testing (e.g., unit, feature, integration, system, acceptance),

b) types of tests to be performed (e.g., functional, boundary, usability, performance, regression, interoperability),

c) traceability to requirements,

d) test environment (e.g., relevancy to customer environment, operational use),

e) test coverage (degree to which a test verifies a product's functions, sometimes expressed as a percent of functions tested),

f) expected results,

g) data definition and database requirements,

h) set of tests, test cases (inputs, outputs, test criteria), and documented test procedure(s),

i) use of external testing, and

j) method of reporting and resolving defects.

7.3.1.C.3-NOTE 1:* Testing may be covered at several levels.

Rationale: Test planning activities ensure that appropriate robust tests are identified. The planning takes into consideration interdependencies of various modules and related tests, identifies problem-reporting procedures, and enforces implementation of time frames for problem resolution depending on the severity of the problem. The test plan is a "blueprint" of the testing activities and identifies all the inputs and outputs required to test the thoroughness of the product.

---

*Notes are suggestions only, not requirements.

TIPS: Remember, this is a "should" requirement and may be replaced by an equivalent process that meets the intent of the requirement.

For each test phase, have a test specification and test plan. Test specifications contain details of tests, including test scenarios. Test plans contain scheduling and resource commitments.

Items (a) through (c) require that the organization document the types and scopes of testing and how they relate to the project requirements. Item (d) requires that the organization document the project impact on customer environment and use. Items (e) and (f) require that the organization document test coverage and results. Item (g) requires that the organization identify specific software items to be included in the database. Items (h) through (j) require that the organization document test procedures, their results, and methods of reporting defects.

NOTE 1 is intended to point out that test planning should be carried out for various levels such as testing of algorithms, modules, subsystems, frames, and so on.

Adder: 7.3.2.C.1 Customer and Supplier Input—The organization shall establish and maintain methods for soliciting and using customer and supplier input during the development of new or revised product requirements.

Rationale: This section of the standard emphasizes establishment of a formal process by the organization to collect and implement inputs from customers and suppliers. The inputs may be related to the development methodology, new features, or changes to the existing product. A structured process of collecting and addressing the input ensures that all customer/supplier concerns and ideas are documented and addressed.

TIPS: This adder requires that the organization contact customers and suppliers for their inputs on new or revised products. A linkage should be established to adders 5.4.2.C.2 (Customer Input) and 5.4.2.C.3 (Supplier Input).

Adder: 7.3.2.C.2 Design and Development Requirements—Design and development requirements shall be defined and documented, and should include:

a) quality and reliability requirements,

b) functions and capabilities of the product,

c) business, organizational, and user requirements,

d) safety, environmental, and security requirements,

e) installability, usability, and maintainability requirements,

f) design constraints, and

g) testing requirements.

Rationale: This adder guarantees that design requirements such as reliability, availability, and serviceability are taken into consideration in the early stages of development. Safety and regulatory requirements are also considered at the design stage. Design limitations must be identified. Documented design requirements facilitate ease in implementing future changes.

TIPS: This is a "should" requirement and may be replaced by an equivalent process that meets the intent of the requirement. Depending upon the type of product or service, the following requirements documents may be needed. They can be addressed in a single document:

- High-level requirements
- Design requirements
- Architectural requirements
- Testing requirements
- Installation
- User documentation
- Performance requirements
- Packaging requirements
- Tolerance requirements
- Safety requirements
- Security requirements
- Maintainability requirements

The following list is relative to installation services only:

a. Service requirements such as crimped connections, soldering, torque, or appearance

b. Service capabilities such as 24/7 support, training output, or technical support

c. User requirements such as installation interval, personnel, tools, or training

d. Safety requirements such as electrical, chemical, or personal

e. Usability requirements such as product- or service-affecting

f. Constraint requirements such as service protection, tooling, or environmental

g. Testing requirements such as test sets or instructions

7.3.6.C-NOTE 1:* Organizations should include customer(s) or a third party during various validation stages, as appropriate.

TIPS: This note points out a means of improving communication with the customer and within the supply chain.

Adder: 7.3.7.C.1 Change Management Process—The organization shall establish and maintain a process to ensure that all requirements and design changes, which may arise at any time during the product life cycle, are managed and tracked in a systematic and timely manner and do not adversely affect quality, reliability, and design intent. Management of changes should include:

a) impact analysis,

b) planning,

c) implementation,

d) testing,

e) documentation,

f) communication, and

g) review and approval.

Rationale: No matter how well the original requirements are collected, documented, and reviewed, changes are unavoidable. Some causes are technology transition, rapid growth, changing business needs, constantly changing customer needs, and competition. This adder ensures that the organization establishes a process in which all the changes to a project's contractual items are documented and managed.

---

*Notes are suggestions only, not requirements.

TIPS: This is a "should" requirement and may be replaced by an equivalent process that meets the intent of the requirement. Corrective action procedures tracks all changes. For hardware, use the product change notice procedures to document and track changes. There are different levels of changes, such as class A or D.[7]

Changes to service requirements must be planned and implemented using a systematic method, such as first office application or laboratory trials. The impact of the change should be documented and provided to users of the service.

One process for change management consists of three steps:

1. *Initiate change request:* A change can be requested by any staff member of the project, by the customer, or by the supplier.
2. *Change request review:* All changes are reviewed to see overall project impact in regards to resources, cost, and schedule. If the impact is low, the change is implemented. If the impact is high, detailed review is conducted to assess the impact.
3. *Implementation of approved changes:* If approved, a requested change is implemented. This may require changing the hardware, software, service, and related documentation or processes.

Change management entails keeping track of changes by documenting the dates when the changes were initiated. The documentation includes description, reason, and impact of the changes. The process of change management may be handled by the maintenance organization, which assures that everyone involved is notified of any changes and keeps abreast of associated risks and dependencies.

Adder: 7.3.7.C.2 Informing Customers—The organization shall establish and maintain a documented procedure(s) to ensure that customers are informed when design changes affect contractual commitments.

Rationale: The contract is a binding document between the organization and the customer. The organization must have an established, documented process to notify the customer of changes that may affect contractual conditions. Notifying customers of any adverse affects on contractual obligations will help manage customer expectations and satisfaction.

TIPS: You may use the Telcordia-recommended procedure for product change notices to satisfy this requirement.[7]

Changes to service methods performed by customers, such as self-installation, must include impact on intervals, tools, training, and so forth, and be provided to customers using the service.

7.4 Purchasing

7.4.1.C.1
Purchasing
procedure(s)

Adder: 7.4.1.C.1 Purchasing Procedure(s)—The documented purchasing procedure(s) shall include:

a) product requirements definition,

b) risk analysis and management,

c) qualification criteria,

d) contract definition,

e) proprietary, usage, ownership, warranty, and licensing rights are satisfied,

f) future support for the product is planned,

g) ongoing supply-base management and monitoring,

h) supplier selection criteria,

i) supplier re-evaluation, and

k) feedback to key suppliers based on data analysis of supplier performance.

7.4.1.C.1-NOTE 1:* The documented procedure(s) should be applicable to off-the-shelf product. This typically includes original equipment manufacturer (OEM) products used in manufacturing and commercial off-the-shelf (COTS) products used in software systems.

Rationale: There has been a large increase in the use of outsourcing of components, assemblies, and services in recent years. Because of this, it is necessary to specify requirements that cover problem areas identified with outsourced products.

TIPS: Items (a) through (f) must be covered in a contract. Items (g) through (j) require the purchasing organization to gather ongoing supplier

*Notes are suggestions only, not requirements.

evaluation data that track the supplier's historical performance, such as tools, test sets, product installability, documentation, training, installation work, and so on, and provide feedback to the supplier. Item (b), risk analysis and management, is especially crucial because of the pressures today on the supply chain. Often many competitors use the same suppliers, which may affect their ability to meet schedules.

For processes that are outsourced, 4.1, General requirements, states that "the organization shall ensure control over such processes."[8] Make sure that the supplier contracts cover details such as the requirements for conducting a "packaging and labeling audit" as required in adder 7.5.5.HS.1.

7.5 Product and service provision

7.5.1.C.1 Organization's support program	*7.5.1.C.2 Service resources*	*7.5.5.C.1 Anti-static protection*

Adder: 7.5.1.C.1 Organization's Support Program—The organization shall ensure that customers are provided support to resolve product related problems.

Rationale: To ensure customer satisfaction, it is important for the organization to designate resources to assist the customer in product-related problems or questions. The support personnel shall be easily available to ensure that any inquiry regarding a product, customer concern, or problem is satisfactorily addressed. The support program will ensure that a quality plan is established to resolve product-related problems within an acceptable time frame. If the customer and organization are in two different time zones, consideration should be given by the organization to the support hours in order to provide support during mutually acceptable time frames.

TIPS: Some suggestions are: (1) a 24/7 call center, (2) a Web site with "frequently asked questions," and (3) a chat room for individual products. This requirement may be handled by customer and product support organizations, which provide expertise on an as-needed basis. This adder requires technical support be available to fix customer services complaints.

Adder: 7.5.1.C.2 Service Resources—The organization shall provide customer contact employees with appropriate tools, training, and resources necessary to provide effective and timely customer service.

Rationale: To ensure that the customer is effectively supported, the experience, knowledge, and skill set of customer support individuals at the organization's site is essential. Training, tools, and documented processes will equip the support staff to do their job in a timely and efficient manner.

The customer may need support in:

- Installation problems
- Product-related issues
- Delivery-related questions
- Invoice-related questions

TIPS: This adder is applicable to maintenance and field support–type functions, not traditional central office installations. Special courses may be developed and given to customer contact personnel.

Adder: 7.5.5.C.1 Anti-Static Protection—Anti-static protection shall be employed, where applicable, for components and products susceptible to electrostatic discharge (ESD) damage. Consider components and products such as: integrated circuits, printed wiring board assemblies, magnetic tapes and/or disks, and other media used for software or data storage.

Rationale: Equipment such as microprocessors, printed board assemblies, and magnetic storage media can be damaged by ESD. Taking precautions will prevent damage in the engineering lab, the manufacturing or distribution centers, or at the customer site.

TIPS: Establish a general operating procedure for ensuring that protection is provided while handling ESD-sensitive components, assemblies, and equipment. The main tools of protection are wrist straps, floor mats, table mats, and anti-static wrapping materials.

8.2 Monitoring and measurement

8.2.1.C.1 Customer satisfaction data	*8.2.3.C.1 Process measurement*

Adder: 8.2.1.C.1 Customer Satisfaction Data—The organization shall establish and maintain a method to collect data directly from customers concerning their satisfaction with provided products. The organization shall also collect customer data on how well the organization meets commitments and its responsiveness to customer feedback and needs. This data shall be collected and analyzed. Trends of the data shall be kept.

Rationale: Customer satisfaction is key to the long-term survival of any business. Monitoring customer satisfaction of the quality and support of the products will enable the organization to take immediate corrective actions in the event that customer satisfaction starts slipping. Delay in responding to customer questions, issues, or problems can affect customer satisfaction, resulting in loss of business and customers.

TIPS: Adverse impacts should be monitored and managed. A quick turnaround time for resolving problems or getting back to the customer is important.

Adder: 8.2.3.C.1 Process Measurement—Process measurements shall be developed, documented, and monitored at appropriate points to ensure continued suitability and promote increased effectiveness of processes.

Rationale: The relationship between the quality of a process and the quality of the product is multifaceted. This relationship depends on many factors, such as robustness of the processes, employees who monitor the processes, and tools that are used to accomplish the processes. When product quality deteriorates, it is difficult to determine which one of the three factors was the biggest contributor. For this reason, process measurements are kept to give accurate and meaningful information. The measurements provide data for improving the reliability of the product and improving the processes used to develop the product. The purpose of the measurements is to determine tangible data information for managing, estimating, and sizing projects or products.

TIPS: This adder requires that process measurements be taken and monitored to ensure process effectiveness.

8.4 Analysis of data

8.4.C.1
Trend analysis of
nonconforming product

Adder: 8.4.C.1 Trend Analysis of Nonconforming Product—Trend analysis of discrepancies found in nonconforming product shall be performed on a defined, regular basis and results utilized as input for corrective and preventive action.

Rationale: Controlling nonconforming products is a factor in continual improvement. The history of nonconforming products must be kept and the data must be utilized to conduct root cause analysis (RCA). The results of the RCA activities are used to implement processes that would correct the nonconformance and also to prevent future occurrences. Trend analysis gives a barometer of the implemented processes over time to see if they are yielding desired quality results.

TIPS: This adder requires that trend analysis be performed on discrepancies at regular intervals and fed to the preventive/corrective action process. Use documented procedures for doing measurement analysis. Standardized reports should be reviewed on a regular basis for corrective action. One organization did not have a database for tracking corrective actions. A process was put in place to track the corrective actions on the ISO 9000 database.

8.5 Improvement

8.5.1.C.1 Quality improvement program	*8.5.1.C.2 Employee participation*

Adder: 8.5.1.C.1 Quality Improvement Program—The organization shall establish and maintain a documented Quality Improvement Program to improve:

a) customer satisfaction,

b) quality and reliability of the product, and

c) other processes/products/services used within the organization.

Rationale: Subsection 8.5.1, Continual improvement, requires the organization to perform a continuous PDCA activity on the quality management system. Adder 8.5.1.C.1 focuses the improvement process on customer satisfaction, product, and processes. Periodically evaluating the results of the quality improvement program shows whether the program is bringing the desired results. If the results are sagging, this is an indication to review the entire quality improvement program so effective changes can be implemented.

TIPS: This adder requires that there be a documented quality improvement program. The responsibility for ensuring the effectiveness of this program belongs to top management. The key mechanism for accomplishing this is management review.

8.5.1.C-NOTE 2:* Inputs to the continual improvement process may include lessons learned from past experience, lessons learned from previous projects, analysis of measurements and post-project reviews, and comparisons with industry best practices.

TIPS: This note suggests using lessons learned from past experience. The organization needs methods of gathering and validating data. The difficulty in using customer satisfaction data is translating the inputs into actions. This can be done by linking this data to the corrective and preventive action process.

Adder: 8.5.1.C.2 Employee Participation—The organization shall implement methods for encouraging employee participation in the continual improvement process.

Rationale: The livelihood of any company depends on the employees' knowledge and experience and the quality of the products. Employees have a direct impact on the way a product is built and serviced. Their participation in identifying the problem areas of the product and processes should result in root cause analysis and preventive—as well as corrective—actions. Encouraging employees to voice concerns regarding deviations from preestablished processes or making them "owners" of the quality of the products can result in better products and improved employee morale.

TIPS: This adder requires that employees have an opportunity to participate in the quality improvement process. One company uses a kaizen program and monitors the number of suggestions submitted. Another method is the use of quality circles. Reports and charts can be published on customer feedback.

8.5.2.C-NOTE 1:* Undesirable deviations from plans and objectives are considered nonconformances.

TIPS: This note is meant as an aid in determining if a condition is considered a nonconformance.

---

*Notes are suggestions only, not requirements.

8.5.2.C-NOTE 2:* Review of corrective action is intended to ensure that the action taken was effective. Review activities may include ensuring that the root cause was properly identified and addressed, appropriate containment action was taken, and corrective actions have not introduced additional problems.

TIPS: This note suggests that a review of corrective action implementation should include assuring that new problems have not been introduced.

8.5.2.C-NOTE 3:* Consideration should be given to include training as part of implementing corrective and preventive actions.

TIPS: This note suggests that the preventive/corrective action solutions could be used to provide training.

ENDNOTES

1. ANSI/ISO/ASQ Q9001-2000, *Quality Management Systems—Requirements*, 3rd ed. (Milwaukee: ASQ Quality Press, 2000).
2. The QuEST Forum, *TL 9000 Quality Management System Requirements Handbook*, Release 3.0 (Milwaukee: ASQ Quality Press, 2001).
3. The QuEST Forum, *TL 9000 Quality Management System Measurements Handbook*, Release 3.0 (Milwaukee: ASQ Quality Press, 2001).
4. Alka Jarvis and Vern Crandall, *Inroads To Software Quality* (Englewood Cliffs, NJ: Prentice Hall, 1997).
5. GR-1202-CORE, *Generic Requirements for a Customer Sensitive Quality Infrastructure*, Issue 1 (Morristown, NJ: Telcordia Technologies, 1995).
6. SR-3535, *Bellcore CSQP℠ Program*, Issue 1 (Morristown, NJ: Telcordia Technologies, 1995).
7. GR-209-CORE, *Generic Requirements for Product Change Notices*, Issue 3 (Morristown, NJ: Telcordia Technologies, 1998).
8. The QuEST Forum, *TL 9000 Quality Management System Requirements Handbook*, Release 3.0 (Milwaukee: ASQ Quality Press, 2001): 4-1.

*Notes are suggestions only, not requirements.

3

Hardware, Software, and Service Requirements

INTRODUCTION

As mentioned in the preface, the TL 9000 requirements are based on ANSI/ISO/ASQ Q9001-2000.[1] Of the 81 "adders," 42 are hardware, software, service, or a combination of two types. In this chapter we repeat the requirements from the *Requirements Handbook,*[2] give a rationale for each requirement, and provide "tips" from organizations that have implemented TL 9000. The reader should focus on the tips because they provide real implementation information.

The standard definition for the word *shall* in ANSI/ISO/ASQ Q9001-2000 is used in TL 9000, but *should* has a different meaning. The word *shall* is used for mandatory requirements that must be adhered to; *should,* on the other hand, indicates "the preferred approach and organizations choosing other approaches must be able to show that their approach meets the intent of TL 9000."[2]

CODES USED TO IDENTIFY THE TYPE OF REQUIREMENT

Individual codes were given to identify each adder as hardware-, software-, service-specific, or when the adder is common to hardware, service, or software:

C: Common (hardware, software, and service)

HS: Hardware and software

HV: Hardware and service

H: Hardware only

S: Software only

V: Service only

Table 3.1 identifies all the adders and corresponding ANSI/ISO/ASQ Q9001-2000 elements and indicates when the adder applies to a specific area, such as software, hardware, service, or combinations of two. There are a total of 42 adders in these categories.

HARDWARE/SOFTWARE ADDERS

TL 9000 includes adders that are common to both software and hardware design and development. We identify these adders in the next few pages, give you a rationale as to why they are a value-add to a quality system, and provide tips on implementing them.

Table 3.1 Adders.

ANSI/ISO/ASQ Q9001-2000 Element	Software	Hardware	Service	Hardware & Service	Hardware & Software
6.2.2				1	
7.1	3		1		1
7.2.3		1			
7.3.1	2				
7.3.2	2	1			
7.3.3	1		1		
7.3.6	1				
7.3.7		1			1
7.5.1	3		2		2
7.5.2				1	
7.5.3		2			1
7.5.5	1	1			1
7.6		1			
8.2.4	1	4		2	
8.4		1	1		
8.5.2	1				
Total	**15**	**12**	**5**	**4**	**6**

7 Product realization

7.1.HS.1 Configuration management plan	7.3.7.HS.1 Problem resolution configuration management	7.5.1.HS.1 Emergency service
7.5.1.HS.2 Installation plan	7.5.3.HS.1 Product identification	7.5.5.HS.1 Packaging and labeling audit

Adder: 7.1.HS.1 Configuration Management Plan—The organization shall establish and maintain a configuration management plan, which should include:

a) identification and scope of the configuration management activities,

b) schedule for performing these activities,

c) configuration management tools,

d) configuration management methods and documented procedure(s),

e) organizations and responsibilities assigned to them,

f) level of required control for each configuration item, and

g) point at which items are brought under configuration management.

7.1.HS.1-NOTE 1:* General work instructions defining general configuration management tasks and responsibilities need not be replicated as part of a specific documented configuration management plan.

Rationale: Configuration management (CM) controls information describing the physical and logical characteristics of resources, and the relationships between those resources. Well-defined CM is a key component to the delivery of any product or service and assists in identifying version number, part number, or documentation. CM gives important information, such as the versions of products supported in the field and the versions of products that an individual customer may have. There are several automated tools in the industry to assist in the CM activity.

TIPS: This is a "should" requirement and may be replaced by an equivalent process that meets the intent of the requirement. A documented

*Notes are suggestions only, not requirements.

CM process will identify the responsibilities for the CM activity, resources used, rules for CM control, the tools used, and a schedule and list of all deliverables/activities. For software the process should cover builds, sanity checks, and testing procedures.

Adder: 7.3.7.HS.1 Problem Resolution Configuration Management— The organization shall establish an interface between problem resolution and configuration management to ensure that fixes to problems are incorporated in future revisions.

Rationale: This activity ensures that fixes in the current product are also incorporated in future revisions so the same errors do not plague the new revisions.

TIPS: In one company, configuration management tools and problem reporting tools are integrated. Review boards assure that problem solutions are carried forward in future versions.

Adder: 7.5.1.HS.1 Emergency Service—The organization shall ensure that services and resources are available to support recovery from emergency failures of product in the field throughout its expected life.

Rationale: Emergency service is a key factor in creating customer confidence and satisfaction. The supplier needs to allocate adequate resources throughout the product lifecycle to assist the customer in case of emergencies.

TIPS: In one company the repair services organization has a procedure for supporting the customer based on the Telcordia CSQP[SM3,4] procedures.

Adder: 7.5.1.HS.2 Installation Plan—The organization shall establish and maintain a documented installation plan. The installation plan shall identify the resources, the information required, and list the sequence of events. The results associated with the installation sequence shall be documented.

Rationale: Documented procedures facilitate quick installation and support. Identification up front of installation schedule, resources, tools, and risks/mitigation avoids delays. Documenting "lessons learned" and providing this knowledge throughout the organization will avoid similar problems in future installations and will assist in root cause analysis activities.

TIPS: Consider using elements of the design project plan (adder 7.3.1.C.1) in developing this plan.

Adder: 7.5.3.HS.1 Product Identification—The organization shall establish and maintain a process for the identification of each product and the level of required control. For each product and its versions, the following shall be identified where they exist:

a) product documentation,

b) development or production tools essential to repeat product creation,

c) interfaces to other products, and

d) software and hardware environment.

Rationale: Design documents and architectural drawings are extremely important when a software or hardware product has to be re-created. In case of a component failure, product identification will make it easier to replace the component. The traceability of product and versions also allows easy tracking of nonconforming hardware or software and facilitates root cause analysis. When a product is recalled due to severe defects, it is usually recalled by identifying the part number. A missing part identification will create a massive problem at recall time or at warranty fulfillment time.

TIPS: Coordinate the hardware aspects of this requirement with the hardware traceability requirements (adders 7.5.3.H.1 and 7.5.3.H.2).

Adder: 7.5.5.HS.1 Packaging and Labeling Audit—The organization shall include a packaging and labeling audit in the quality plan or documented procedure(s). This may include, for example, marking, labeling, kitting, documentation, customer-specific marking, and correct quantities.

7.5.5.HS.1-NOTE 1:* This audit is normally done on products ready to ship.

Rationale: The documented audit procedure will ensure effectiveness of the process related to packaging and labeling. Adequate packaging will prevent damage caused in shipping or handling. Labeling ensures correct identification of the product for installation, warranty, maintenance, recall, or repair.

TIPS: Difficulty may arise with fulfilling this requirement if an organization outsources some of its processes. The requirement covering outsourcing of processes in Section 4.1, General requirements, requires the organization to "ensure control over such processes."[1] If the supplier provides product directly to the customer or to the organization in final packaged form, the organization must include a packaging and labeling audit at the supplier's location in its procedures.

*Notes are suggestions only, not requirements.

HARDWARE AND SERVICE ADDERS

6 Resource management

6.2.2.HV.1
Operator qualification

Adder: 6.2.2.HV.1 Operator Qualification—The organization shall establish operator qualification and requalification requirements for all applicable processes. These requirements, at a minimum, shall address employee education, experience, training, and demonstrated skills. The organization shall communicate this information to all affected employees.

Rationale: Untrained, inexperienced, and underqualified employees can negatively impact quality and productivity. The qualification and experience required to perform each job at the supplier's site needs to be documented. This will assist the operators in obtaining state-of-the-art training and remaining current in their knowledge of the job that they are performing. Any knowledge can become obsolete due to changing technology; therefore requalification is required to remain current.

TIPS: Consider education, experience, training, and demonstrated skills when identifying operator qualifications. For installation service, this adder requires installer qualification and requalification criteria be established and communicated to employees.

7 Product realization

7.5.2.HV.1
Operational changes

Adder: 7.5.2.HV.1 Operational Changes—Each time a significant change is made in the established operation (e.g., a new operator, new machine, or new technique), a critical examination shall be made of the first unit(s)/service(s) processed after the change.

Rationale: When significant changes are made, it is necessary to assure that the product is not affected. The process of thorough testing and examination will reveal whether the change is free of defects that would negatively impact the product.

TIPS: Records should be kept of when the change was made, the reason for the change, the impact, and the name of the individual who implemented the change. Changes must be truly significant. For example, if a process is automated, an operator change may not impact the quality of the output and a "first article" inspection may not apply. For installation service, a new installer, tool, or method must result in the examination of the initial outputs.

8 Measurement, analysis and improvement

8.2.4.HV.1 Inspection and test documentation	*8.2.4.HV.2 Inspection and test records*

Adder: 8.2.4.HV.1 Inspection and Test Documentation—Each inspection or testing activity shall have detailed documentation. Details should include the following:

a) parameters to be checked with acceptable tolerances,

b) the use of statistical techniques, control charts, etc.,

c) sampling plan, including frequency, sample size, and acceptance criteria,

d) handling of nonconformances,

e) data to be recorded,

f) defect classification scheme,

g) method for designating an inspection item or lot, and

h) electrical, functional, and feature testing.

Rationale: A well-defined test methodology that requires thorough test and inspection records will aid the testing activity. When each defect found in the test cycle is classified based on its impact on the business, the number of defects can be used to determine the release of the product.

TIPS: This is a "should" requirement and may be replaced by an equivalent process that meets the intent of the requirement. The defect data, when analyzed to determine the root cause, can be used to ensure that similar defects are prevented in the future.

Adder: 8.2.4.HV.2 Inspection and Test Records—Inspection and test records shall include:

a) product identification,

b) quantity of product inspected,

c) documented inspection procedure(s) followed,

d) person performing the test and inspection,

e) date of inspection and/or test, and

f) number, type, and severity of defects found.

Rationale: Recording this information provides the supplier with the capability of tracing product from the field and/or repair back to the tests performed. This capability is a valuable tool for improvement of the inspection and testing process and the quality of the product.

TIPS: This information is useful for both production and installation of hardware. It is valuable in developing a culture of communication, feedback, and improvement that links product support, repair, and testing.

HARDWARE ADDERS

7 Product realization

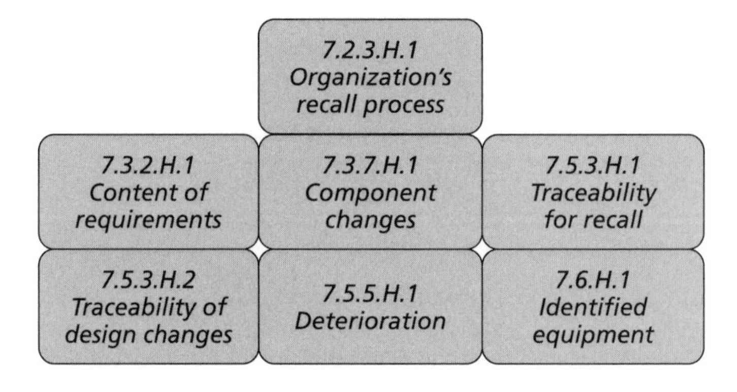

Adder: 7.2.3.H.1 Organization's Recall Process—The organization shall establish and maintain a documented procedure(s) for identifying and recalling products that are unfit to remain in service.

Rationale: Identification of resources to address field failures and support repair or exchange of product at customer sites will reduce the risk to the customer.

TIPS: Coordinate with or link this process to adders 7.5.3.H.1, Traceability for Recall, and 7.5.3.HS.1, Product Identification.

Adder: 7.3.2.H.1 Content of Requirements—The product requirements shall include, but are not limited to:

a) nominal values and tolerances,

b) maintainability needs, and

c) end-item packaging requirements.

Rationale: This element requires consideration of the above features during the design phase and ensures that the product's performance criteria are considered early in the product lifecycle.

TIPS: Major characteristics to be considered are reliability, maintainability, maintenance support, and availability.

Adder: 7.3.7.H.1 Component Changes—The organization shall have a documented procedure(s) in place to ensure that material or component substitutions or changes do not adversely affect product quality or performance. The documented procedure(s) should include:

a) functional testing,

b) qualification testing,

c) stress testing,

d) approved parts listing, and/or

e) critical parts listing.

Rationale: It is very common for hardware components to be replaced by less expensive, equivalent components. When this happens, critical tests and/or analyses must be made to assure that there is no effect on functionality, quality, or reliability.

TIPS: This is a "should" requirement and may be replaced by an equivalent process that meets the intent of the requirement. Use a change management tool to track changes and maintain records that can later be reviewed and analyzed. This requirement can be covered as part of the change management process, adder 7.3.7.C.1.

Adder: 7.5.3.H.1 Traceability for Recall—Field Replaceable Units (FRU) shall be traceable throughout the product life cycle in a way that helps organizations and their customers to identify products being recalled, needing to be replaced, or modified.[5]

Rationale: In case of a failure of previously released units, identification of products will assist in recalling the remaining units in a timely manner with less impact. Also, it will enable the supplier to identify those units that can be changed or modified in the field.

TIPS: This requirement can be included under the change management process. Use bar codes, serial numbers, and/or CLEI (Telcordia provided) codes.[6] Product support should keep track of units in the field based on customer inputs.

Adder: 7.5.3.H.2 Traceability of Design Changes—The organization shall establish and maintain a documented procedure(s) which provides traceability of design changes to identifiable manufacturing dates, lots, or serial numbers.

Rationale: Documenting design changes with dates, serial numbers, and lots will assist in recalling units in a timely manner with less impact. Also, it will enable the supplier to identify those units that can be changed or modified in the field. This information will facilitate easy withdrawal or identification of field units when a defect is detected due to design changes.

TIPS: This requirement can be included under the change management process, adder 7.3.7.C.1. Use the engineering change order procedure.[5] After approval of the change order, a date is identified when the change will be implemented. Manufacturing maintains serial numbers and tracks them for changes.

Adder: 7.5.5.H.1 Deterioration—Where the possibility of deterioration exists, materials in storage shall be controlled (e.g., date stamped/coded) and materials with expired dates shall be deemed non-conforming.

Rationale: This requirement is aimed at identification of materials that, if outdated, may cause products to malfunction.

TIPS: Identifying such materials by date stamp or code will make it easy to track outdated materials, thereby preventing their use.

Adder: 7.6.H.1 Identified Equipment—Monitoring and measuring devices that are either inactive or unsuitable for use shall be visibly identified and not used. All monitoring and measuring devices that do not require calibration shall be identified.

Rationale: Satisfying this requirement will prevent use of measuring equipment that is not adequate for test or production and will also identify equipment that does not require calibration. The purpose is to avoid the use of equipment that may give inaccurate or wrong results during production.

TIPS: Periodic audits of laboratory equipment is a means of assuring compliance to this requirement.

8 Measurement, analysis and improvement

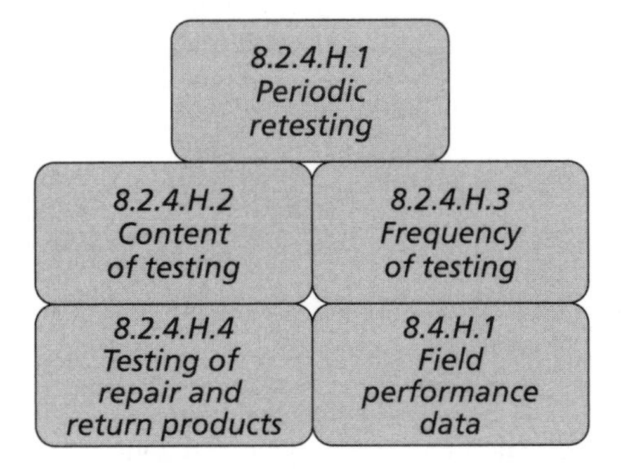

Adder: 8.2.4.H.1 Periodic Retesting—The organization shall establish and maintain a documented procedure(s) that ensures products are periodically retested to assess the product's ability to continue to meet design requirements.

Rationale: Periodic retesting ensures that the product continues to perform according to the design requirements, and any changes made to the original design have not caused the product to deviate from the original design intent.

TIPS: This should be included in the test plan (see adder 7.3.1.C.3). Document a procedure for testing individual units to original requirements and system-level summary tests.

Adder: 8.2.4.H.2 Content of Testing—The initial test and periodic retest shall be more extensive than the routine quality tests. The initial test shall include those that are contained in the customer's and/or organization's product specifications and/or contracts. The results of these tests shall be documented.

8.2.4.H.2-NOTE 1:* Product specifications may include environmental, vibration, flammability, and operational stress type testing.

Rationale: The initial tests will reveal the usability of the product, which will be a determining factor for the release of the product.

TIPS: This should be included in the test plan (see adder 7.3.1.C.3). Document a procedure for testing individual units to original requirements and system-level summary tests.

Vigorous testing of the initial product will uncover a majority of the defects. All the defects observed during testing must be documented, resolved, and maintained as quality records.

Adder: 8.2.4.H.3 Frequency of Testing—The organization shall establish and document the frequency for test and periodic retest. When determining the test frequency, the organization shall include the following:

a) product complexity and service criticality,

b) number of design, engineering and/or manufacturing changes made to the product and whether the change(s) affect form, fit, and/or function,

c) changes to the manufacturing process,

d) manufacturing variations, (e.g., tooling wear),

e) material and/or component substitutions and failure rates, and

f) the field performance record of the product.

---

*Notes are suggestions only, not requirements.

Rationale: For thorough testing of the product, the frequency of tests must be determined on valid criteria, such as product complexity, frequency of changes, and the number of changes to the product.

TIPS: This should be included in the test plan (see adder 7.3.1.C.3). Document a procedure for testing individual units to original requirements and system-level summary tests.

Adder: 8.2.4.H.4 Testing of Repair and Return Products—Repair and return products shall be subjected to the appropriate evaluation(s) and/or test(s) to ensure functionality to product specification.

Rationale: Repaired and returned products are expected to be the equivalent of new products. Therefore, they must go through the same rigorous testing procedures as new products. Failure of a unit a second time will have a bad effect on customer confidence.

TIPS: Ensure that there are no deviations from the documented testing procedures when it comes to returned or repaired products. This will help prevent return of repaired units a second time.

Adder: 8.4.H.1 Field Performance Data—The quality management system shall include the collection and analysis of field performance data which can be used to help identify the cause and frequency of equipment failure. In addition, no trouble found (NTF) data shall also be maintained. This information shall be provided to the appropriate organizations to foster continual improvement.

Rationale: The data analysis of failures found in the field and identification of root cause of the failures will allow the implementation of preventive and corrective actions, and should result in changes to processes to prevent similar failures in the future. Timely response to customer complaints and inquiries will result in improved customer satisfaction. No trouble found (NTF) units are of special concern because they cause a loss of confidence in the diagnostic systems.

TIPS: Set up a database to gather field failure data and to link into a similar database containing repair and NTF data. Mechanisms should be set up to track NTF units returned to the field in order to prevent repetitive returns on the same board. If repair is done by a third party, the supplier or the customer should require repair and NTF data be made available to the design organizations.

SOFTWARE ADDERS

7.1 Planning of product realization

7.1.S.1 Estimation	7.1.S.2 Computer resources	7.1.S.3 Support software and tools management

Adder: 7.1.S.1 Estimation—The organization shall establish and maintain a method for estimating and tracking project factors during project planning, execution, and change management.

Rationale: In project management, sound estimation techniques are crucial to the timely completion of the project. Poor estimation can result in cost and schedule overruns. Constant changes to the original product requirements will impact the schedule and cost. If the changes are not managed, the quality of the product will be affected.

TIPS: The estimation process should be ongoing during the design and development process. Use electronic tools to estimate resource needs throughout the project. Use inputs from other projects. Gather a body of knowledge.

7.1.S.1-NOTE 1:* Project factors should include product size, complexity, effort, staffing, schedules, cost, quality, reliability, and productivity.

TIPS: The estimation process should take into consideration project-related factors such as product size, complexity, effort, staffing, schedules, cost, quality, reliability, and productivity. You may want to perform a code complexity evaluation to estimate testing time, and also conduct risk analysis up front so you can mitigate some of the risks that may delay the project and affect estimation.

Adder: 7.1.S.2 Computer Resources—The organization shall establish and maintain methods for estimating and tracking critical computer resources for the target computer, the computer on which the software is intended to operate. Examples of these resources are utilization of memory, throughput, real time performance, and I/O channels.

Rationale: The reliability, availability, and performance of the software depends on a thorough understanding of the resources needed by the

*Notes are suggestions only, not requirements.

deployed product. It's important during the design phase to continually estimate the utilization of memory, throughput, real-time performance, and I/O channels in the target computer.

TIPS: The same tools used in adder 7.1.S.1 may be used to estimate the resources needed by the target computer. Use inputs from other projects. Gather a body of knowledge.

Adder: 7.1.S.3 Support Software and Tools Management—The organization shall ensure that internally developed support software and tools used in the product life cycle are subject to the appropriate quality method(s). Tools to be considered include: design and development tools, testing tools, configuration management tools, and documentation tools.

Rationale: If the internally-developed tools do not go through a rigorous validation, verification, and calibration process, then they may adversely affect the quality of the software.

TIPS: Use the same development and test procedures for design software and development tools that you use for development of production software. Place scripts for testing and problem tracking under configuration control.

7.3 Design and development

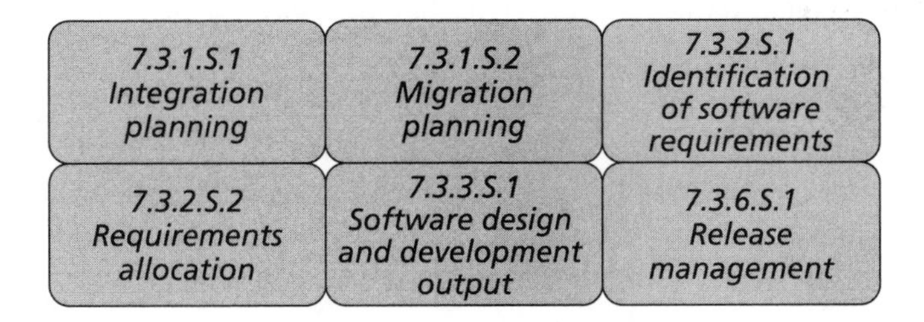

Adder: 7.3.1.S.1 Integration Planning—The organization shall develop and document a plan to integrate the software components into the product to ensure they interact as designed. The plan shall include:

a) methods and documented procedure(s),

b) responsibilities,

c) schedule for integration, and

d) test requirements.

Rationale: Once all the modules (programs) are validated individually, the modules should be integrated and tested as a "whole system." When integrating the modules together a number of problems can occur: one module can have an adverse effect on another; data can be lost across interfaces or when modules are combined; or the combined modules may not produce the expected results. Integration testing will uncover the above mentioned "glitches" when the modules are combined together and tested to detect errors resulting from the integration.

TIPS: Use an integration test plan template that covers responsibilities, environmental requirements, and other elements of planning requirements (see adder 7.3.1.C.3).

Adder: 7.3.1.S.2 Migration Planning—The organization shall develop and document a migration plan when a system or software product is planned to be migrated from an old to a new environment. The plan should include the following:

a) requirements analysis and definition of migration,

b) development of migration tools,

c) conversion of product and data,

d) migration execution,

e) migration verification, and

f) support for the old environment in the future.

Rationale: Migration from an old to a new environment must be documented to ensure that no functions are lost or degraded when an existing system is replaced with a new one. Migration planning should ensure that degradation of the reliability and performance of the software system does not take place.

TIPS: This is a "should" requirement and may be replaced by an equivalent process that meets the intent of the requirement. The migration plan should include components such as requirements analysis, software migration tools, data verification, and maintenance and support of the old environment during the migration process.

Adder: 7.3.2.S.1 Identification of Software Requirements—The organization shall determine, analyze, and document the software component requirements of the system.

Rationale: A thorough requirements document affects the contents and functionality of a product. The requirements should be verified up front to ensure that there is a common understanding of what is needed and how it will be accomplished. This understanding is necessary to both the supplier and the customer.

TIPS: Use requirements document reviews to uncover errors such as unclear requirements, missing functionality, or usability issues. This requirement may be covered as part of adder 7.3.2.C.2, Design and Development Requirements.

Adder: 7.3.2.S.2 Requirements Allocation—The organization shall document the allocation of the product requirements to the product architecture.

Rationale: The management of the requirements during the design and architecture definition guarantees that every requirement is addressed in the development.

TIPS: Requirements allocation will ensure that the requirements are not "lost" as the project progresses from one phase to the next. Documenting the allocation will also help to ensure that appropriate test cases are written for the requirements, therefore giving full test coverage. This requirement may be covered as part of adder 7.3.2.C.2, Design and Development Requirements.

Adder: 7.3.3.S.1 Software Design and Development Output— Software design and development outputs should include, but are not limited to:

a) system architecture,

b) system detailed design,

c) source code, and

d) user documentation.

Rationale: The deliverables from the design—the design outputs— will assist in the maintenance of the software product during its life. These outputs will also facilitate testing and future changes.

TIPS: This is a "should" requirement and may be replaced by an equivalent process that meets the intent of the requirement. Use separate documents to cover architectural design, detailed design, source code, user documentation, and so on. Outputs refer back to requirements in the parent document. One organization uses a template from their requirements planning document.

Adder: 7.3.6.S.1 Release Management—The organization shall establish and maintain a documented procedure(s) to control the release and delivery of software products and documentation. The documented procedure(s) should include methods to provide for the following:

a) release planning information delivered to the customer sufficiently in advance of the release,

b) product introduction and release schedules to the customer,

c) detailed descriptions of product features delivered and changes incorporated in new software products or releases, and

d) advising the customer of current or planned changes.

Rationale: To control the delivery of a product, development of a release plan is necessary. The intent of this document is to describe the contents of the release and the flow of the engineering release process. Entrance and exit criteria for the various stages of the release process are also described.

At the same time, the customer needs time to prepare to receive the new product, allocate resources to support it, and schedule training for the staff. Receiving a detailed content list of the product features will assist the customer in conducting acceptance testing and in ensuring that the product has all the functions that were stated in the requirements document.

TIPS: This is a "should" requirement and may be replaced by an equivalent process that meets the intent of the requirement. The project plan described in adder 7.3.1.C.1 can be used as a guide in planning the release and delivery of software products and documentation.

Some inputs to be considered at the time of release are:

- Product change notices
- Customer information reports
- Beta tests with customers
- Customer documentation tests
- Release notes with exceptions
- Products that go to customer labs first

7.5 Production and service provision

7.5.1.S.1 Patching procedure(s)	7.5.1.S.2 Patch documentation	7.5.1.S.3 Replication	7.5.5.S.1 Software virus protection

Adder: 7.5.1.S.1 Patching Procedure(s)—The organization shall establish and maintain a documented procedure(s) that guide the decision to solve problems by patching.

a) The documented procedure(s) shall address patch development procedures, propagation (forward and backward), and resolution.

b) The documented procedure(s) shall be consistent with customer needs or contractual requirements for maintenance support.

c) The organization shall provide the customer with a statement of impact on the customer's operation for each patch.

Rationale: Often there is a dilemma regarding whether to send a patch to a customer for resolving defects or to send a product update. Depending on how many changes/fixes are addressed in a patch and the impact of the changes/fixes to the customers, the supplier should establish a procedure to identify whether a patch or an update should be sent.

TIPS: See adder 7.5.1.S.2 for related information covering the documentation of patches.

Adder: 7.5.1.S.2 Patch Documentation—The organization shall establish and maintain methods to ensure that all documentation required to describe, test, install, and apply a patch has been verified and delivered with the patch.

Rationale: Patches are necessary to address emergency fixes that are supplied to customers to address specific defects. The supplier must establish a process to test these emergency fixes to make certain that they work and do not create adverse effects on the existing system. A customer's ability to install the patch smoothly without encountering difficulties depends on the process used by the supplier to test the patch prior to the delivery. Verification of patch-related documentation will help ease customer use and assure accuracy.

TIPS: See adder 7.5.1.S.1 for related information.

Adder: 7.5.1.S.3 Replication—The organization shall establish and maintain a documented procedure(s) for replication, which should include the following:

a) identification of master copy,

b) identification of replicate copies for delivery,

c) quantity of replicates to deliver,

d) type of media,

e) labeling,

f) identification of required documentation such as user guides,

g) packaging of documentation, and

h) control of environment to ensure repeatable replication.

Rationale: The customer may have several sites and may require the supplier to create and deliver more than one copy of the product. This adder ensures that there are no defects introduced during labeling, packaging, and duplication. Identification of the master copy will prevent the risk of duplicating from an obsolete copy.

TIPS: This is a "should" requirement and may be replaced by an equivalent process that meets the intent of the requirement. If replication is done at another facility, the design organization should retain a "gold" copy and send a "silver" copy to the replication center. The replication center can then send a sample back to the design organization for validation testing against the gold copy.

Adder: 7.5.5.S.1 Software Virus Protection—The organization shall establish and maintain methods for software virus prevention, detection, and removal from the deliverable product.

Rationale: In today's environment, we have seen the negative results of virus attacks. A virus can bring down entire businesses and result in losses of millions of dollars. In order to prevent the customer from encountering the ill effects of any type of virus, the supplier must establish processes to protect against unwanted penetration by the virus.

TIPS: Develop a procedure for creating disks and CDs and for ensuring that no virus has been added. Test each disk before sending it out.

8 Measurement, analysis and improvement

8.2.4.S.1 Test documentation	*8.5.2.S.1 Problem resolution*

Adder: 8.2.4.S.1 Test Documentation—Software tests shall be conducted according to a documented procedure(s) and the test plan. Documentation of testing shall include:

a) test results,

b) analysis of test results,

c) conformance to expected results, and

d) problem reporting for nonconforming items.

Rationale: Testing is a complex engineering activity and must be planned in order to have effective results. The supplier must identify how the product will be tested, the types of tests to be conducted, and document actual and expected results of the tests. If the expected results were not identified in each test case, how would one know whether the test had passed or failed? Any failed test results must be documented and the failed tests fixed and retested prior to the release of the product.

TIPS: Use a rigorous system test procedure in which test cases are developed from the requirements by an organization independent of the design organization. System testers should be challenged to "break the system." Document system test and validation test plans. Reports should have test results, conformance to expected results, and a list of nonconforming items.

Adder: 8.5.2.S.1 Problem Resolution—The organization shall establish and maintain a documented procedure(s) to initiate corrective action once a reported trouble is diagnosed as a problem. The documented procedure(s) should provide guidelines for distinguishing among potential solutions such as:

a) patching,

b) immediate source code corrections,

c) deferring solutions to a planned release, and

d) providing documented "work-around" operational procedure(s) and resolution within a designated timeframe based on the severity of the issue.

Rationale: A great deal of time is wasted in analyzing and fixing similar problems that continue to plague every release of the software. By having a structured process for conducting root cause analysis and implementing corrective and preventive actions, the supplier may prevent "same type" defects from reoccurring in the future.

TIPS: This is a "should" requirement and may be replaced by an equivalent process that meets the intent of the requirement. This requirement may be satisfied in conjunction with adder 7.5.1.S.1, Patching procedures. The method of addressing defects by using a patch, immediate source code corrections, or deferring the solution to a future planned release is common to satisfying this adder and adder 7.5.1.S.1.

SERVICE ADDERS

7 Product realization

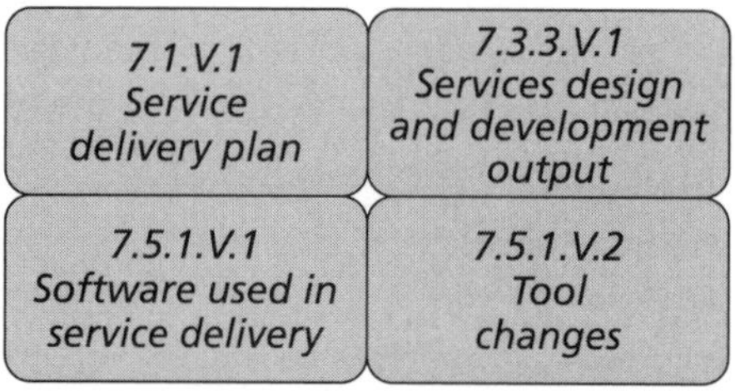

7.1.V.1 Service delivery plan	*7.3.3.V.1 Services design and development output*
7.5.1.V.1 Software used in service delivery	*7.5.1.V.2 Tool changes*

Adder: 7.1.V.1 Service Delivery Plan—Suppliers that are responsible for the delivery or implementation of a service, and are not responsible for the design and development of that service, shall comply with the Project Plan requirements of 7.3.1.C.1.

Rationale: This adder requires use of a project plan as defined in adder 7.3.1.C.1, Project plan. This will assist in controlling schedule slippage and will identify issues before they impact the project adversely.

TIPS: The reference to adder 7.3.1.C.1 makes this is a "should" requirement. It may be replaced by an equivalent process that meets the intent of the requirement. Some of the items in adder 7.3.1.C.1 may not be required for service delivery. Be sure to document why they are not required.

Adder: 7.3.3.V.1 Services Design and Development Output—The required output from the services design and development shall contain a complete and precise statement of the service to be provided. Design and development outputs shall include, but are not limited to:

a) service delivery procedures,

b) resource and skill requirements,

c) reliance on suppliers,

d) service characteristics subject to customer evaluation, and

e) standards of acceptability for each service characteristic.

Rationale: In order to expedite service delivery with minimal impact on the customer, the supplier must have documented agreements with the customers that outline the type and extent of the service to be provided. In addition, the supplier must have trained staff with thorough understanding of the service delivery procedures. The acceptance criteria for the service and the standards of acceptability must be identified and documented as a part of the service contract to avoid any misunderstanding at a later time between the supplier and the customer.

TIPS: The following list of outputs is for installation services only:

a. Development of installation manuals

b. Development of a list of tools, test sets, and installer skills

c. Development of a list of noninstaller operations

d. Development of a list of items (such as support documentation, diagnostic capabilities, and ease of use) to be evaluated by the customer

e. A list of customer functional and performance requirements

Adder: 7.5.1.V.1 Software Used in Service Delivery—Organizations shall document and implement processes for the maintenance and control of software used in service delivery to ensure continued process capability and integrity.

Rationale: Delivery of some services requires the use of software tools. Examples are tools used to download the software being installed and tools to do diagnostics during the installation. These tools can have defects that will affect their ability to deliver the service. Identifying and resolving defects will assist in minimizing service interruptions in the field due to malfunctioning software.

TIPS: The organizations responsible for the service should have a documented procedure for verifying and validating the software used in service delivery. The methodology for verifying and validating the software used in service delivery is similar to the methodology for verifying and validating software delivered as part of a product.

Adder: 7.5.1.V.2 Tool Changes—The organization shall have a documented procedure(s) in place to ensure that substitutions or changes to tools used in performing the service do not adversely affect the quality of the service.

Rationale: Changes to tools used in performing the service can adversely affect the quality of the service.

TIPS: It is important to document a structured process describing how tool changes will be reviewed for their effect on the service. This should include operational testing of the tools before actual use.

8 Measurement, analysis and improvement

8.4.V.1
Service
performance data

Adder: 8.4.V.1 Service Performance Data—The quality management system shall include the collection and analysis of service performance data, which can be used to identify the cause and frequency of service failure. This information shall be provided to the appropriate organizations to foster continual improvement of the service.

Rationale: A quality system is a closed-loop process where measurements must be taken to determine the quality of products, processes, and services. Service failure data should be used for implementing corrective and preventive actions. The performance data gives you information to determine where improvement can be made in the process.

☞ **TIPS:** This adder requires that there be a system in place to collect service-related problems, and that the data be analyzed and fed to the appropriate organizations for use in quality improvement.

ENDNOTES

1. ANSI/ISO/ASQ Q9001-2000, *Quality Management Systems—Requirements*, 3rd ed. (Milwaukee: ASQ Quality press, 2000).
2. The QuEST Forum, *TL 9000 Quality Management System Requirements Handbook*, Release 3.0 (Milwaukee: ASQ Quality Press, 2001).
3. GR-1202-CORE, *Generic Requirements for a Customer Sensitive Quality Infrastructure*, Issue 1 (Morristown, NJ: Telcordia Technologies, 1995).
4. SR-3535, *Bellcore CSQP^SM Program*, Issue 1 (Morristown, NJ: Telcordia Technologies, 1995).
5. GR-209-CORE, *Generic Requirements for Product Change Notices*, Issue 3 (Morristown, NJ: Telecordia Technologies, 1998).
6. Common Language Equipment Codes. Visit the Web site at www.telecordia.com/resources/commonlanguage/product .

4

TL 9000 Measurements

THE IMPORTANCE OF MEASUREMENTS

One of the fundamental drivers for creating the QuEST Forum and TL 9000 was the telecommunication industry's inability to measure* the impact of improvements on the cost of quality. Both service providers and suppliers were helpless in this regard. Another motive was to measure in real terms the effect of applying the TL 9000 requirements and metrics to the telecom industry. The *Measurements Handbook* is intended to help guide progress and evaluate results of the TL 9000 quality system implementation.

Another major driver behind measurement was the following: *Unless we know where we are, there is no way we can define goals and objectives for future improvement.* With the solid push for continuous improvement, it was imperative that TL 9000 adopt a measurements standard.

All business excellence criteria place extensive emphasis on measurements. The Malcolm Baldrige National Quality Award (MBNQA) Criteria for Performance Excellence[1] states: ". . . *measures* . . . represent factors that lead to improved customer, operational, and financial performance. A comprehensive set of *measures* . . . represents a clear basis for aligning activities with the company's goals." And ". . . (a balanced *set of measures*) . . . offers an effective means to communicate . . . priorities, to monitor actual performance, and to marshal support for improving results."

**Measurements* and *metrics* are used interchangeably in chapters 4 and 5. In many cases, metrics is used because it is the term that was used during the development of the *TL 9000 Metrics Handbook*, Release 2.5.[7]

The European Foundation for Quality Management (EFQM) model[2] states the importance of measurements as "*Management by Facts . . .*" and "[the] EFQM model is a practical tool to help organizations do this by *measuring* where they are on path to Excellence."

There can be no evaluation without metrics. For example, any type of sport activity is meaningless without some kind of measurement. For some reason, there is a negative connotation attached to the term *metrics*. Some people think that the sole purpose of having metrics is to compare their performance against their peers or competitors. People fail to realize the broader important perspective behind metrics collection.

In the telecommunication environment, in particular, metrics would be beneficial in order to:

- Determine whether stated goals and expected benefits of the TL 9000 quality system requirements are being achieved
- Protect integrity and enhancement of value of telecommunications products and services
- Achieve continuous improvement of products and subscriber services
- Enhance customer–supplier relationships
- Improve subcontractors' performance
- See overall cost reduction and increased competitiveness of suppliers and service providers
- Show the value of quality and reliability successes of TL 9000 participants, and encourage wider TL 9000 implementation and acceptance
- Help direct quality system management activity to areas where the most beneficial results will be realized
- Show all industry stakeholders, such as customers, employees, stockholders, suppliers, and the public, how their interests are being addressed
- Determine performance and quality relative to others in the industry

INTRODUCTION

The uniqueness of the TL 9000 quality standard lies in its requirement for metrics submission. No other known quality standard in any industry has a similar requirement. The main purpose for metrics submission is to make available a few vital measurements by which suppliers can judge their own performance in relation to the industry norm. A long-term purpose is to create some type of yardstick for the telecom industry by which the industry as a whole can measure its progress.

As anticipated, initially there was some resistance towards this move for measurements. But it soon wilted because everyone realized that the improvements cannot be measured without the metrics and metrics can only help everyone in the industry.

METRICS WORK GROUP

The next question was how to go about defining which metrics to include in the handbook. A team of quality experts with background in measurements was chartered to define the metrics that would be part of the handbook. The team set out to perform their new assignment with enthusiasm. They soon realized the enormity of the task in hand and decided to hold monthly meetings for the creation of the metrics handbook.

Initially the discussions were lengthy and sometimes heated, focusing on what type of metrics should be considered. The choice was between product metrics, process metrics, or both. The service providers wanted both product and process metrics to be included. On the other hand, the suppliers justifiably argued that if too many metrics were included, the cost of collecting, reporting, and monitoring them would be very high and would eventually derail the overall goal of cost reduction.

After the debates lasted for a few sessions, the metrics work group approached the handbook creation process very professionally. To avoid further arguments, the team first decided to agree on rules and guidelines for the development of TL 9000 metrics.

Based on these guidelines, a list of criteria was developed that would make the metrics package acceptable. The criteria were:

1. Well-defined counting rules for each measurement
2. Product applicability rules based on a well-defined, unambiguous, product category table*
3. Clear, auditable, collection and reporting processes

*The product category table can be viewed on the QuEST Forum Web site: www.questforum.org.

4. Unambiguous answers to any questions that may arise with respect to:
   - Logistics associated with suppliers collecting data from each subset of their customers
   - Logistics surrounding supplier data aggregation
   - Logistics regarding a geographically dispersed supplier, where not all metrics for a given product can be reported by a given geographic organization
   - Logistics regarding the relation between all suppliers in the supply chains and their role in metrics collection/reporting
   - Logistics and industry ramifications regarding published benchmark data

The criteria were effectively used in the development of the metrics handbook.

Metrics Ground Rules

1. The metrics can fall into the following categories: (a) hardware; (b) software; (c) services; or (d) common (to hardware, software, and services).
2. Each measure has a defined audience and purpose, and is consistent, objective, easily understood, and useful for evaluation or decision making.
3. Each measure provides data to support timely analysis to determine trends, projections, or cause and effect.
4. Use common basis (normalized) for similar products/services—making inevitable comparisons as fair as possible.
5. Select a small number per category.
6. Use or modify existing metrics where possible—for example, service provider report cards, RQMS (GR 929-CORE),[3] IPQM (GR-1315-CORE),[4] Supplier Data (GR-1323-CORE).[5]
7. Define the metrics as potentially part of a possible future industry index (similar to J.D. Power).
8. Address different product types (for example, cable, switch, outside plant hardware, switch systems/software, cross-connect systems/software, and so on) and define for all phase(s) of their lifecycles.

METRICS SELECTION USING THE GOAL–QUESTION–METRIC MODEL

The metrics work group decided to adopt the goal–question–metric (GQM) model to determine which metrics to select. The GQM paradigm explains a measurement and evaluation method[6] that entails asking questions to help identify goals. For each goal, ask questions that will help one understand whether the goal has been accomplished. Many questions have answers that can be measured.

The GQM paradigm is a useful tool in identifying good, appropriate metrics. It entails three aspects:

1. What are the goals of metrics?
2. What questions would customers and suppliers like answered by these metrics?
3. What are the metrics that would answer these questions?

The GQM exercise led to the following list of goals:

Goals of Metrics

1. Reduce the cost of ownership
   1A. By the purchasers
   1B. By the suppliers
2. Increase customer satisfaction and confidence
3. Increase predictability (on-time delivery, delivered functionality, cost, final product quality)
4. Improve current performance (product and process):
   4A. Product related
   4B. Process related
5. Improve lifecycle performance
   5A. Process related
   5B. Product related
6. Reduce cost impact of software faults (supplier and purchaser) as identified in 1A and 1B
7. Improve product quality throughout its lifecycle
8. Reduce cycle times for software delivery

The next step involved creating a set of questions for each goal. The following list of questions is a sample to demonstrate the approach taken by the group. Sample questions are provided for goals 1A, 1B, 2, 5B, and 8.

In view of the enormous task of defining metrics, the group decided to develop the metrics in two parts. Phase 1 would include field metrics, or the metrics collected after a product is delivered and installed in the field. Phase 2 would include process metrics, which would exhibit the development process efficiency and effectiveness.

Sample Questions

1A. Reduce the cost of ownership by the purchaser

- How much does it cost the purchaser?
  - To acquire
  - To install/implement
  - To operate (day to day—normal conditions)
  - To maintain (corrective—cost of failure)
  - To grow (adaptive)
  - To retire

1B. Reduce the cost of ownership by the supplier

- How much does it cost the supplier?
  - To acquire/develop/enhance (concept through delivery)
  - To manufacture/deploy/install
  - To support (day to day under normal conditions)

2. Increase customer satisfaction and confidence

- How satisfied are our customers?
- What factors do our customers find important?
  - On-time delivery
  - Product performance
  - Product support

continued

continued

- Product introduction
- Technology leadership (perception)
- How satisfied and confident are our customers with:
  - Our ability to meet commitments
  - The quality of our products and services
  - Our responsiveness to customer requests
  - The fix response time (responsiveness)

5B. Improve released product quality (service provider)

- How many field failures were found by severity?
- How many field outages occurred (events and duration)?
- Did the delivered products meet the requirements?
  - Feature functionality
  - Interoperability
- What is the "out-of-the-box" quality (installability, initial operation, documentation, completeness of delivery, perhaps a system issue)?
- What areas of the software have the highest concentration of defects?
- What is the quality of the software fixes?

8. Reduce cycle times for software delivery

- How much time is spent in each development phase and activity?
- What are the bottlenecks in the supplier's process?
- What is the overall cycle time release to release?
- How much functionality is delivered within a given cycle window (production rate)?
- How effectively are resources being planned and used?

PHASE 1 MEASUREMENTS

The list of measurements from Phase 1 is divided into five groups: a) common: applicable to hardware, software, and service categories; b) hardware and software: applicable to both hardware and software only; c) hardware: applicable to hardware products only; d) software: applicable to software products only; and e) service: applicable to service category only. The Phase 1 measurements are listed in the exhibit below. The detailed descriptions of metric counting rules, calculations, formulas, and example calculations are provided in the *Measurements Handbook*.[7]

All suppliers are expected to establish a measurement management system through which metrics collection, monitoring, and reporting takes place. The unique aspect of this system is that service providers also have a major role to play, because for some measurements service provider cooperation is a must. Unless they provide field data to the suppliers, the suppliers cannot calculate certain measurements. The measurements model in Figure 4.1 depicts, at a higher level, the cooperation between suppliers, service providers, and the QuEST Forum.

Figure 4.1 shows the flow and usage of TL 9000 measurements as described in the handbook.[7] It explains the interaction between service

Phase 1 Measurements

- Common
  1. Number of problem reports
  2. Problem report fix response time
  3. Overdue fix responsiveness
  4. On-time delivery
- Hardware and software
  1. System outage
- Hardware
  1. Return rates
- Software
  1. Software installation and maintenance
     a. Release application aborts
     b. Corrective patch quality
     c. Feature patch quality
     d. Software update quality
- Service
  1. Service quality

providers, suppliers, and the metrics repository system (MRS). The three must work together to make measurements collection, reporting, and analysis a reality. In addition, contractually required measurements are included to provide the complete picture as to how measurements will be used to drive continuous improvement in the telecommunications industry.

Explanation of Figure 4.1

1. Service provider must provide the supplier with data for some measurements
2. The supplier calculates the measurements and provides them to the measurements administrator (UTD)
3. UTD uses mathematical tools (3A) to calculate statistics (3B) by product class
4. UTD calculations are posted on the QuEST Forum Web site
5. Inputs are provided to the QuEST Forum about continuous improvement programs from external data (2), QuEST Forum Web site (4), internal TL 9000 measurements (5A), and assessment report/report cards (5B)
6. Service provider may require other data from the supplier

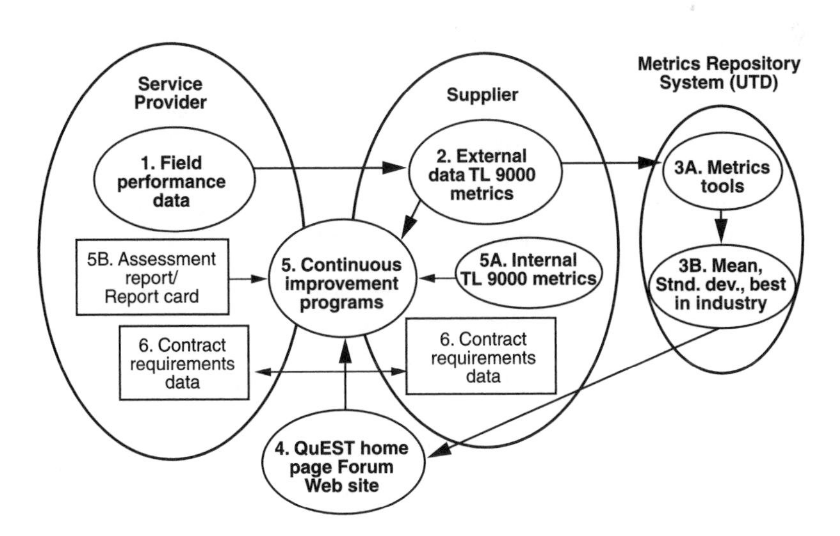

Figure 4.1 QuEST TL 9000 measurements model.

METRICS USAGE

There are three primary uses of the metrics:

1. Reporting to Metrics Repository System (MRS)
2. Internal use for continuous improvement programs
3. To strengthen customer–supplier relationships

The following principles for processing and usage of the metrics are meant to be consistent with the purpose of providing an environment where service providers and suppliers can work together to drive continuous improvement:

- Suppliers will provide TL 9000 metrics to the Forum Administrator (UTD) who will compile all the metric data and calculate an "industry mean," "standard deviation," and "best in industry" for each category. Results or reports produced by the UTD will not identify individual suppliers.
- Service providers may request their suppliers to provide the TL 9000 metrics specific to that service provider. This information exchange will take place strictly between the supplier and the service provider on a contractual basis. The Forum and metric administrators will not be involved in any way.
- There will not be ranking by the measurements administrator.

The general uses of the measurements currently envisioned are:

- "Industry mean," "standard deviation," and "best in industry" statistics will be used by suppliers to improve products and services. The measurements provide data that is needed by suppliers to identify significant improvement areas. After improvement actions have been taken, the data will allow suppliers to determine the level of improvement and decide if additional significant improvement is needed.
- "Industry mean," "standard deviation," and "best in industry" statistics, along with supplier-specific metrics, may be used by service providers to evaluate and work with suppliers. By comparing a supplier's performance against the industry data, gaps in performance can be determined and the service provider and supplier can address closing the gaps.
- Suppliers and service providers will use the metrics to determine if products, intercompany processes, and services meet end-user expectations.

- Improvement of industry products and services, based on QuEST Forum initiatives and goals.
- Harmonization of the supplier assessment process.

After the process and measurements have matured:

- They will be used to establish a Telecommunication Quality Index to be published worldwide
- The Forum may create an "award program" for best in class

The following text provides an overview of the Phase 1 measurements. All measurements are product- or service-related.

COMMON MEASUREMENTS

Common measurements are applicable to the hardware, software, and service categories. There are four common measurements: (1) number of problem reports; (2) problem report fix response time; (3) overdue fix responsiveness; and (4) on-time delivery.

Number of Problem Reports

The number of problem reports (NPR) is computed as the total number of problem reports per normalized unit per month. This measurement is adopted from RQMS[3] and applies to all products.

The purpose of this measurement is to evaluate the number of problem reports or complaints during field operations, in order to reduce their number along with associated costs and revenue losses. NPR provides a common means whereby customers and suppliers can evaluate the rate of problem reports, including engineering complaints (EC), which are reported on the supplier's product. This should induce joint customer/supplier efforts to reduce costs and revenue losses associated with problems.

Monitoring the trends of problem reports helps the customer understand whether product quality in the field is improving. It is a measure of the reported field "pain" experienced by the customer.[3]

Only customer-reported problem reports are counted. Both customers and suppliers provide data for NPR measurement. Since the origin of this metric is in the RQMS,[3] an alternative way of reporting is also described in the handbook.

One report is to be compiled per supplier product category.[8] Typical target audiences for NPR data are customer maintenance, engineering,

Problem Reports per System—Six Month Moving Average

	Mar-99	Apr-99	May-99	Jun-99	Jul-99	Aug-99	Sep-99	Oct-99	Nov-99	Dec-99	Jan-00	Feb-00	Mar-00
Critical	0.000	0.000	0.000	0.000	0.040	0.033	0.033	0.033	0.033	0.033	0.000	0.017	0.017
Major	0.000	0.000	0.033	0.025	0.060	0.100	0.117	0.117	0.133	0.133	0.100	0.067	0.067
Minor	0.200	0.250	0.333	0.375	0.440	0.450	0.550	0.650	0.700	0.650	0.550	0.500	0.417
Systems in Service 100's	10.00	12.50	15.00	13.75	11.60	10.33	11.71	11.00	10.33	9.60	9.27	9.17	9.38

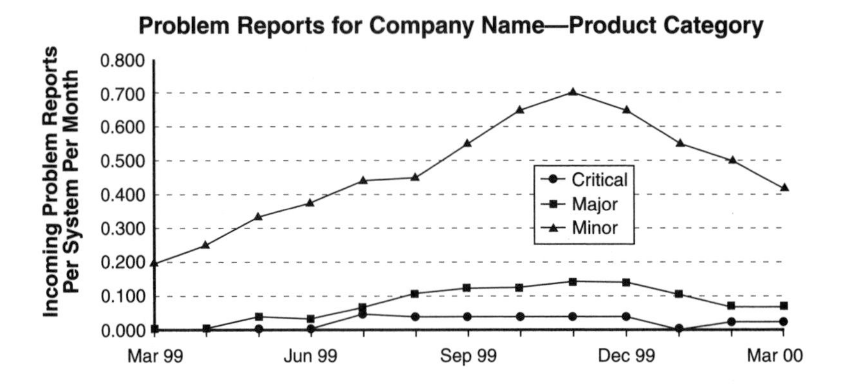

Figure 4.2 Total problem reports by severity level.

technical support, supplier technical support, manufacturing management, and engineering management.

The measurement to be reported to UTD must be in a tabular form as described in the handbook. A graphical form depicting the trend of minor, major, and critical problem reports is illustrated in Figure 4.2.

Problem Report Fix Response Time and Overdue Problem Report Fix Responsiveness

Although the *Measurements Handbook* lists problem report fix response time (FRT) and overdue problem report fix responsiveness (OFR) separately for the ease of reading and application, we are discussing them together in this section due to their inherent similarity.

FRT represents the responsiveness of suppliers to a subset of customer-reported problems that require a change to the product. FRT measures the overall responsiveness of the supplier to all major and minor problems reported in hardware and software categories and all problems reported under the service category. OFR measures the rate of closure of overdue major and minor hardware and software problems and all service problem reports. The rules of NPR and FRT apply to OFR. Both measurements are adopted from RQMS.[3]

The FRT measurement for each product category includes the following:

- Percent hardware/software major problems fixed within 30 days
- Percent hardware/software minor problems fixed within 180 days
- Percent service problems fixed on time

The intent is to get the suppliers to respond to problem reports in an efficient and effective manner, so as to facilitate prompt availability of end-user services. Critical problem reports are excluded because they are worked exclusively until their resolution.

The FRT measurement monitors the supplier's responsiveness to problem reports that, for resolution, may require changes to the product: hardware fixes; document changes; procedure changes; and/or software fixes (patches, S/W upgrades, or code changes in future generations). The calculation measures the percentage of fixes due to be made that meet the closure targets.

The overdue fix responsiveness calculates percentages from the number of problem reports (PRs) overdue, the number of penalty PRs (for non-service problems) and the number of overdue PRs closed each month for both major and minor severity reports.

The data for this measurement is provided by the service providers. Figures 4.3 and 4.4 are examples of the overall fix response time and overdue fix responsiveness. Figure 4.3 includes data on critical problems.

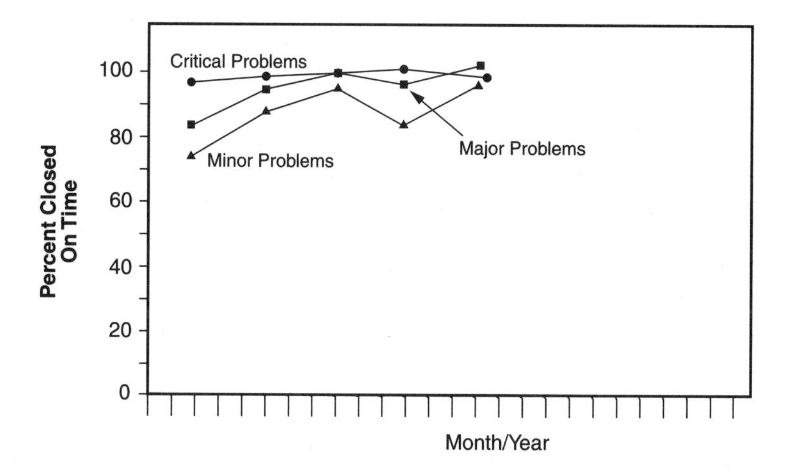

Figure 4.3 Overall fix response time.

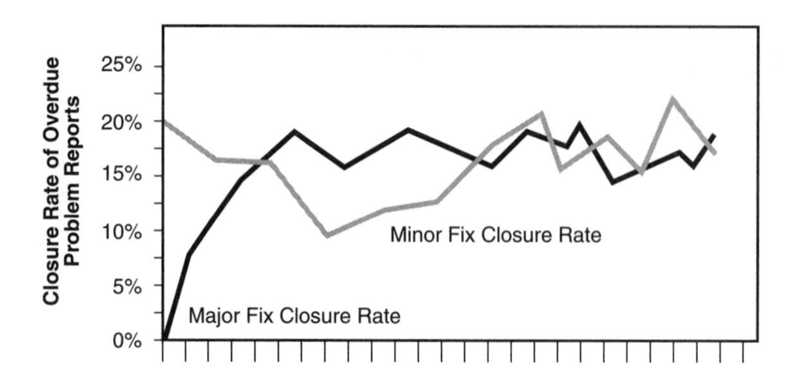

Figure 4.4 Overdue fix responsiveness.

On-Time Delivery

The purpose of measuring on-time delivery (OTD) is to evaluate the supplier's performance in order to meet the customer's need for timely product and service delivery and to meet end-user expectations.

There are three measurement identifiers for OTD: (1) OTIS (on-time installed system delivery); (2) OTI (on-time items delivery); and (3) OTS (on-time service delivery). OTD is measured against customer requested date (CRD) for each of three distinct cases (systems, items, and services).

- OTIS: *installed system order fulfillment.* OTD for installed systems is computed as follows:
  Percent (%) systems accepted on CRD = (# systems accepted on CRD) ÷ (# systems for which CRD occurred during the reporting period)
- OTI: *orderable line item order fulfillment.* OTD for orderable line items is computed as follows:
  Percent (%) orderable line items delivered on CRD = (# of orderable line items delivered complete on CRD) ÷ (# of orderable line items for which CRD occurred during the reporting period)
- OTS: *service order fulfillment.* OTD for service orders is computed as follows:
  Percent (%) service orders delivered on CRD = (# of service orders delivered complete on CRD) ÷ (# of service orders for which CRD occurred during the reporting period)

Tables 4.1 and 4.2 illustrate computation of OTD metrics from a series of orderable line items (Table 4.1) and installations of systems (Table 4.2).

Table 4.1 Example of orderable line item OTD.

	PO	Line Item	Qty	CRD	Qty	Actual	OTD CRD	NOTE
	A	1	5	3/10	5	3/10	1	
		2	6	3/12	6	3/13	0	
		3	4	3/17	4	3/18	0	
	B	1	8	3/20	8	3/22	0	
		2	12	3/22	6	3/22	0	3
					6	3/25		
		3	2	3/29	2	?	0	5
		4	2	3/30	2	3/30	0	4
	C	1	7	2/15	7	3/15	NA	2
		2	1	2/15	1	3/15	NA	2
TOTAL	**3**	**9**	**47**	**7**			**1**	
March		**OTD**					**14%**	**1**

Notes:

1. The CRD OTD performance for March was 14%, or 1 (CRD met)/7 (CRDs due).
2. PO line items C1 and C2 CRDs were not counted in the total of 7 for March because they had February CRD dates.
3. PO line item B2 was not on time for CRD because only ½ of the items were delivered to CRD.
4. PO line item B4 was given no credit for partial on time.
5. PO line item B3 has "?" because the actual delivery date since this item was not delivered by the end of March when this report was prepared.

Table 4.2 Example of installed system OTD.

	PO	CRD	Line Item	Qty	Actual	Acceptance	OTD CRD	Note
	A	3/10	1	5	3/10		1	
			2	6	3/10			
			3	4	3/10	3/10		
	B	3/20	1	8	3/22		0	
			2	12	3/22			
					3/25	3/25		
	C	3/21	1	2	3/21		0	
			2	2	3/21			
					3/22	3/22		
	D	2/15	1	7	3/15		0	2
			2	1	3/15	3/15		
TOTAL	**4**	**3**	**9**	**47**			**1**	
March	**OTD**						**33%**	**1**

Notes:

1. The CRD-installed OTD performance for March was 33%, or 1 (CRD met) ÷ 3 (CRDs due).
2. PO system D CRDs were not counted in the total of 3 for March, because it had a February CRD date.

	Date	1996	1997	Jan	Feb	Mar	Apr	May	Jun
TYPE	Met CRD	15	22	3	2	1	2	2	3
Systems	Total	40	45	5	4	3	5	4	5
	CRD	38%	49%	60%	50%	33%	40%	50%	60%
	LOR	6	5	1	0	0	0	0	0
Line Items	Met CRD	18	22	2	3	1	3	4	4
	Total	45	50	5	6	7	7	8	7
	CRD	40%	44%	40%	50%	14%	43%	50%	57%
	LOR	8	6	0	1	0	0	0	0

LOR = Late Order Received

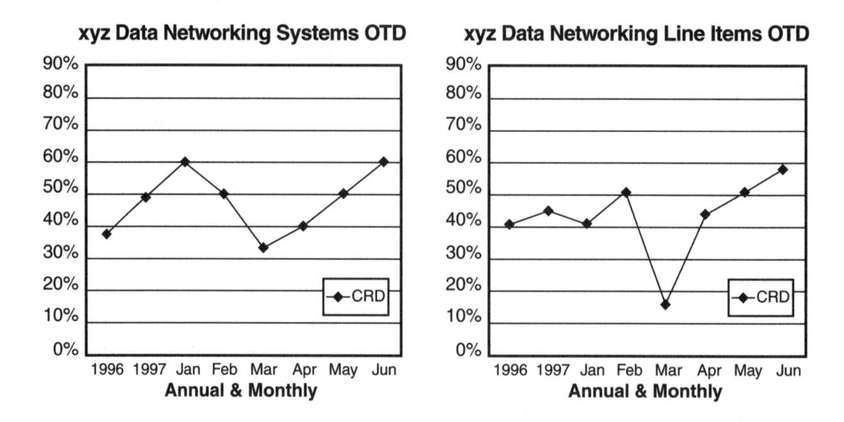

Figure 4.5 Example charts and graphs for on-time delivery metrics.

Reports of OTD data may be presented in a table and chart as shown in Figure 4.5. The acronyms used in the tables and figure are customer requested date (CRD), purchase order (PO), and quantity (Qty).

HARDWARE AND SOFTWARE MEASUREMENT

System Outage

The purpose of measuring system outage is to evaluate the downtime performance and outage frequency during field operation in order to reduce both the frequency and duration of outages and their associated cost and revenue impacts.

System outage applies to hardware and software products and is a measure of complete loss of functionality of all or part of a system. The system outage downtime measurement varies by product. The system outage frequency measurement is outages per normalization unit per year.

The system outage measurement is computed for the following aspects:

- Annualized outage frequency
- Annualized downtime
- Annualized supplier attributable outage frequency
- Annualized supplier attributable downtime

An example of system outage is presented in Table 4.3. It contains an example of overall TL 9000 system downtime reporting in an end office.

Consider a population of four systems comprising 1600 terminations distributed as 100, 200, 300, and 1000 terminations per system. The 100- and 1000-termination systems are host systems and each experienced one 10-minute total outage during the quarter. The 200- and 300-termination switches are remote systems. The 200-termination system incurred a 20-minute partial outage affecting 50 terminations. The calculation is given below for average switch size = 400 terminations:

Table 4.3 Example: system outage—switching systems.

Outage Length (minutes)	Terminations Affected	Equivalent Switch Factor	Number of Outages	Normalized Frequency Factor* (Number/ Switches)	Normalized Time Factor* (Minutes/ Switches)
10	100	100/400= 0.25	1	0.250	2.5
10	1000	1000/400= 2.50	1	2.500	25.0
20	50	50/400= 0.125	1	0.125	2.5
Total Quarter			M=3	2.875	30.0
Annualized			12	11.50	120.0

Dividing by the population of 4 switches, the normalized numbers are:

Per Switch Per Year	2.875 Outages (weighted*)	30 Minutes
Per Switch Per Year	3.000 Outages (unweighted**)	

* Weighted outages indicate average outage frequency for a system of typical size.

** Unweighted outages indicate average outage frequency for any system without regard to size.

Hence, a typical system (or a typical line) will experience 2.875 outages totaling 30.0 minutes in a year based on performance in the current quarter. The metric is independent of whether the outage is total or partial and whether the termination is served by a host or remote system. For systems with incremental recovery, the total duration and frequency can be calculated as the sum of the partial weighted outages.

THE HARDWARE-ONLY MEASUREMENT

There is one metric that is applicable to hardware only: return rate.

Return Rate

The return rate applies to field-replaceable hardware products only, for example, circuit packs. However, bulk items such as cable, optical fiber, and mechanical hardware are exempt from this measurement.

The purpose of collecting this metric is threefold: 1) to provide a measure of quality of the product initially received by the customer and during subsequent in-service operation, 2) to determine areas that need corrective action or will most likely benefit from continuous improvement activity, and 3) to provide input data needed to calculate equipment lifecycle costs.

Return rates are measured for hardware products as follows:

- Initial return rate in percentage returns
- One-year return rate in percent returns
- Long-term return rate in percent returns

The one-year year return rate provides an early indication of the in-service performance of both new and existing products. For new products, it provides an indication of the expected performance in the absence of any historical data. For existing products, it provides a means of comparing the expected long-term performance with actual performance when that information becomes available.

The return rate metric is computed for each product category in three parts: (1) initial return rate (IRR), (2) one-year return rate (YRR), and (3) long-term return rate (LTR). The initial return rate (IRR) is the rate of returns for six months prior to the reporting period. The one-year return rate (YRR) is the rate of return of product in its first year of service life following the IRR, that is, for months seven through 18 after shipment. The long-term return rate (LTR) is for months 19 and later. A return rate calculation example is given as follows.[7]

The return rates for January 1999 are calculated as:

$$\text{Initial return rate} = \frac{(\text{Returns for units shipped Jul-98 through Jan-99}) \times 12 \times 100}{\text{Total Shipments for Jul-98 through Dec-98}}$$

$$= \frac{(14 + 16 + 20 + 39 + 36 + 23 + 5) \times 12 \times 100}{8833 + 8954 + 9368 + 9818 + 9787 + 10528}$$

$$= 3.20\%$$

$$\text{One-year return rate} = \frac{(\text{Returns for units shipped Jul-97 through Jun-98}) \times 12 \times 100}{\text{Total Shipments for Jul-97 through Jun-98}}$$

$$= \frac{(22+11+17+19+16+24+11+7+14+10+6+6) \times 12 \times 100}{(8253+9243+9261+9721+10131+10140+6263+6436+7244+7275+7396+8263)}$$

$$= 1{,}96\%$$

$$\text{Long-term return rate} = \frac{(\text{Returns from shipments prior to Jul-97}) \times 12 \times 100}{\text{Total Shipments prior to Jul-97}}$$

$$= \frac{39 \times 12 \times 100}{30000}$$

$$= 1.56\%$$

The return rates for all months in 1998 are plotted in Figure 4.6.

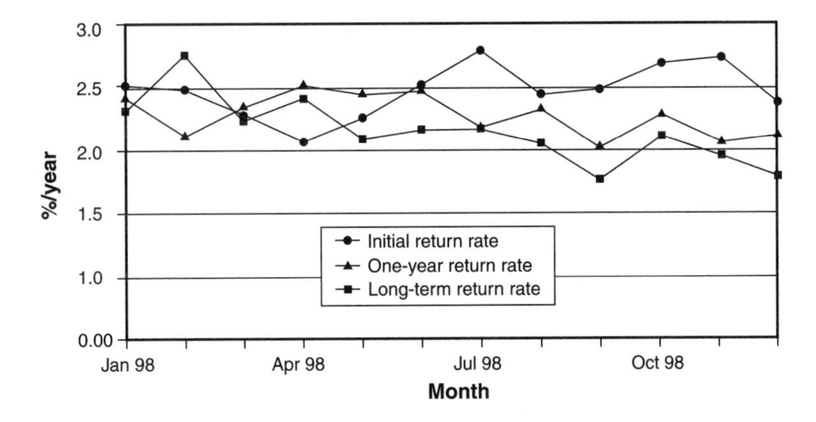

Figure 4.6 1998 Return rates.

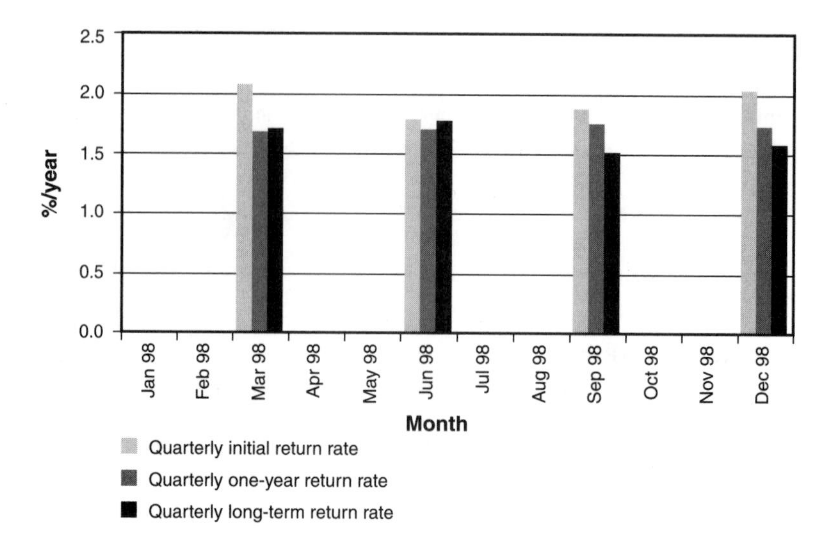

Figure 4.7 Quarterly return rates.

To calculate the rates for the first quarter (Jan-98 through Mar-98), add the return totals and shipment totals for the three months. For example:

$$1\text{Q98 initial return rate} = \frac{(53 + 60 + 49) \times 12 \times 100}{(30133 + 30959 + 32493)}$$

$$= 2.08\%$$

The quarterly return rates for 1998 are plotted in Figure 4.7

SOFTWARE MEASUREMENTS

There are three software measurement options: (1) software release application and patching, (2) software updates and, (3) software update and patching. Their purpose is to track software installation of new releases and the ongoing maintenance effort.

The four essential elements of software measurements are: software updates (SWU), release application aborts (RAA), corrective patch quality (CPQ), and feature patch quality (FPQ). The method used to maintain the software in the field is used to determine which of these measurements are reported.

Option 1—Software Release Application and Patching

This option is for products that use software release application for installation and patching for maintenance.

The following software measurements apply to option 1:

- Release application aborts (RAA)
- Corrective patch quality (CPQ)
- Feature patch quality (FPQ)

Option 2—Software Updates

This option is for products that use software updates for both installation and maintenance. Software updates (SWU) is the only software measurement applicable to option 2.

Option 3—Software Update and Patching

This option is for products that use software updates for installation, and software updates and patches for maintenance.

The following software measurements apply to option 3:

- Software updates (SWU)
- Corrective patch quality (CPQ)
- Feature patch quality (FPQ)

A Description of the Four Software Measurements

The next sections contain a description of each of the four software measurements.

Software Update Quality

A software update is a set of changes to a release and is commonly referred as a *dot* or *point* release. It differs from a patch in the manner in which software changes are made to a system. A software update replaces existing product with a complete new load. On the other hand, a patch replaces a subset of software in the current load without changing the rest of the load.

The SWU metric tracks the percentage of defective software updates. It is the cumulative percent of defective updates for each of the three latest, most dominant releases per product category. The purpose of tracking this metric is to evaluate the level of defective software updates in order to minimize associated customer risk when implementing software updates. The measurements also show the overall trend of defective software updates by release over a 12-month period.

The intent of tracking this measurement is to ensure that suppliers of software carry out the requisite testing of their software (for example, regression tests or other methods). The authors have observed that some software companies, in order to provide a quick solution to a problem in the field, provide an update without undergoing sufficient testing to ensure that the fix does not break something else in the system. The RQMS[3] contains an alternate method of reporting.

Figure 4.8 is an example of a software update metric. The number of software updates is tracked per month for each release. Also, the number of defective software updates is tracked per month for each release. Both monthly and cumulative numbers are currently reported to the MRS, however only the cumulative numbers are used for industry calculations.

Release Application Aborts

The release application abort (RAA) measurement is the percentage of release applications that abort. It measures the percentage of systems with aborts for each of the last three most dominant releases per product category. This metric is derived from RQMS,[3] and RQMS data are an alternative method of reporting the metric.

The purpose of RAA is to minimize the service provider's risk of aborts when applying a software release. The audience includes customer personnel that use the product and/or measure supplier performance. The audience also includes supplier personnel who are involved in the software design/delivery processes, as well as those involved in monitoring/managing the vendor quality system.

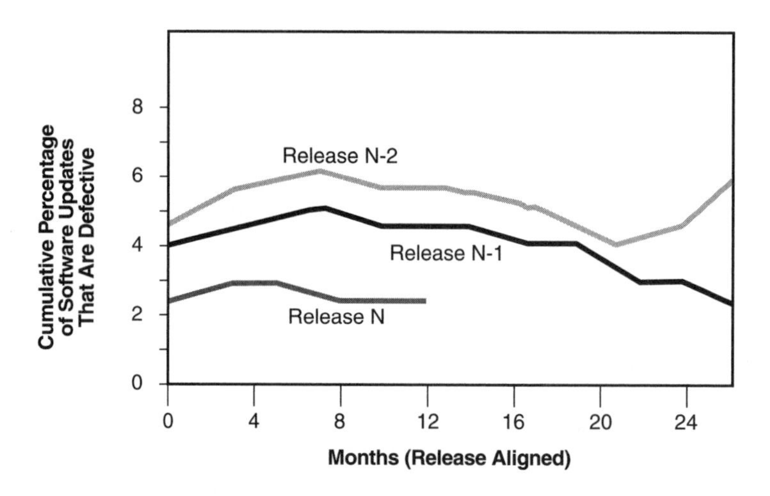

Figure 4.8 Software quality update example.

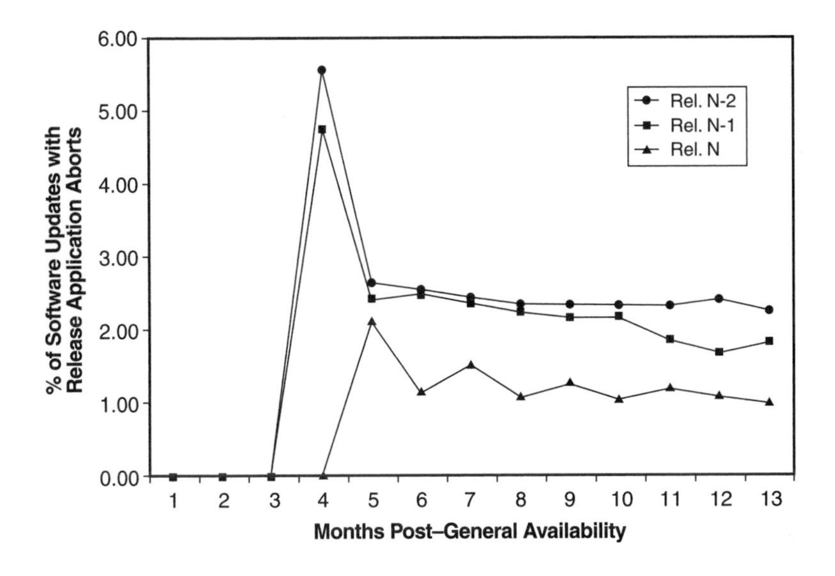

Figure 4.9 Release application abort measurement.

Table 4.4 is an example of RAA. A vendor is upgrading three active releases (from prior releases to release N, N-1, and N-2). Software upgrades numbers and those encountering release application aborts are shown.

The corresponding release application abort plot is shown in Figure 4.9.

Corrective Patch Quality and Feature Patch Quality

A feature patch adds functionality, while a corrective patch fixes a known problem. Errors may occur for each of these types of patches.

Corrective patch quality is the percentage of official corrective patches that are determined to be defective. Feature patch quality is defined as the percentage of official feature patches that are determined to be defective. The purpose of tracking this metric is to evaluate the percentage of defective official patches in order to minimize service provider risk of failure.

Both corrective patch and feature patch quality measurements are computed for each of the latest three most dominant releases. They are applicable to software products only. These measurements are adapted from RQMS,[3] and RQMS data are a valid reporting alternative. The *Measurements Handbook*[7] describes in detail the counting rules, calculations, formulas, and reporting format. An example of quality corrective patch measurements is shown in Table 4.5.

Table 4.4 Release application aborts example.

Month	1	2	3	4	5	6	7	8	9	10	11	12	13
Number of upgrades in month													
Release N-2	1	2	5	10	20	40	45	46	45	43	45	30	24
Release N-1	1	3	5	12	20	39	46	51	52	45	48	33	29
Release N	1	4	6	14	22	39	45	52	54	50	47	36	30
Cumulative number of upgrades to that release													
Release N-2	1	3	8	18	38	78	123	169	214	257	302	332	356
Release N-1	1	4	9	21	41	80	126	177	229	274	322	355	384
Release N	1	5	11	25	47	86	131	183	237	287	334	370	400
Number of upgrades that encountered release application aborts in the month													
Release N-2	0	0	0	1	0	1	1	1	1	1	1	1	0
Release N-1	0	0	0	1	0	1	1	1	1	1	0	0	1
Release N	0	0	0	0	1	0	1	0	1	0	1	0	0
Cumulative number of upgrades that encountered release application aborts to that release													
Release N-2	0	0	0	1	1	2	3	4	5	6	7	8	8
Release N-1	0	0	0	1	1	2	3	4	5	6	6	6	7
Release N	0	0	0	0	1	1	2	2	3	3	4	4	4
Release application aborts measurement (cumulative %)													
Release N-2	0.00	0.00	0.00	5.56	2.63	2.56	2.44	2.37	2.34	2.33	2.32	2.41	2.25
Release N-1	0.00	0.00	0.00	4.76	2.44	2.50	2.38	2.26	2.18	2.19	1.86	1.69	1.82
Release N	0.00	0.00	0.00	0.00	2.13	1.16	1.53	1.09	1.27	1.05	1.20	1.08	1.00

Table 4.5 Example of corrective patch quality.

A supplier has three active releases (N, N-1, and N-2). Corrective patch distribution and bad corrective patch counts are shown in the table below:

Month	1	2	3	4	5	6	7	8	9	10	11	12	13
Number of corrective patches issued in month													
Release N-2	60	55	50	47	42	35	35	31	32	30	29	27	25
Release N-1	55	55	50	40	36	32	34	36	33	32	26	24	24
Release N	52	53	48	35	34	34	32	30	28	30	25	24	22
Number of defective corrective patches identified in month													
Release N-2	1	0	0	2	0	0	0	1	0	1	1	0	0
Release N-1	1	0	0	1	0	0	1	0	0	0	1	0	1
Release N	0	1	0	0	0	1	0	0	0	0	0	1	0
Defective corrective patch measurements (6-month rolling window— % defective corrective patches)													
Release N-2	1.67	0.87	0.61	1.42	1.18	1.04	0.76	1.25	1.35	0.98	1.56	1.63	1.72
Release N-1	1.82	0.91	0.63	1.00	0.85	0.75	0.81	0.88	0.95	0.49	1.04	1.08	1.14
Release N	0.00	0.95	0.65	0.53	0.45	0.78	0.85	0.47	0.52	0.53	0.56	0.59	0.63

A supplier has three active releases (N, N-1, and N-2). The table above shows how the corrective patch distribution and bad corrective patch counts have been calculated.

The corresponding corrective patch quality measurements plot is shown in Figure 4.10.

The plot shown in Figure 4.11 is an example of defective feature patch quality.

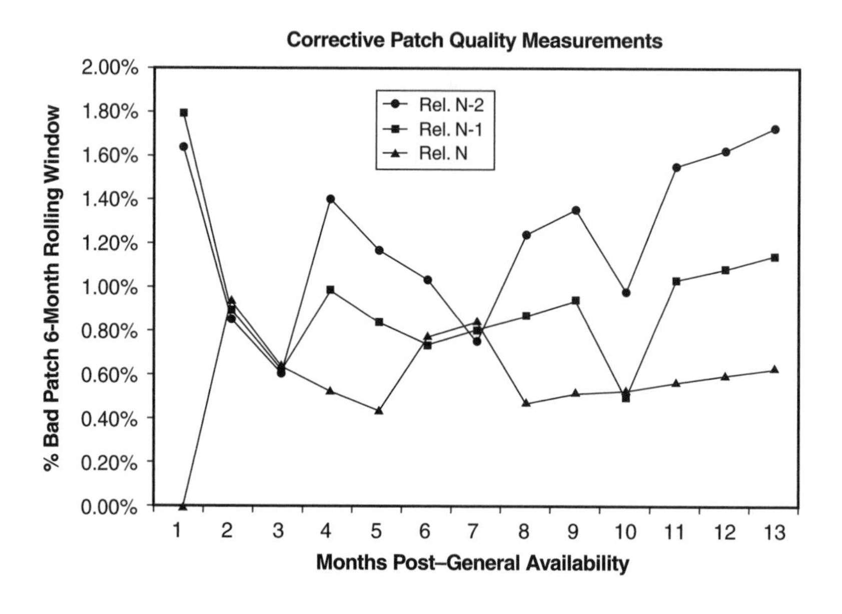

Figure 4.10 Corrective patch quality example.

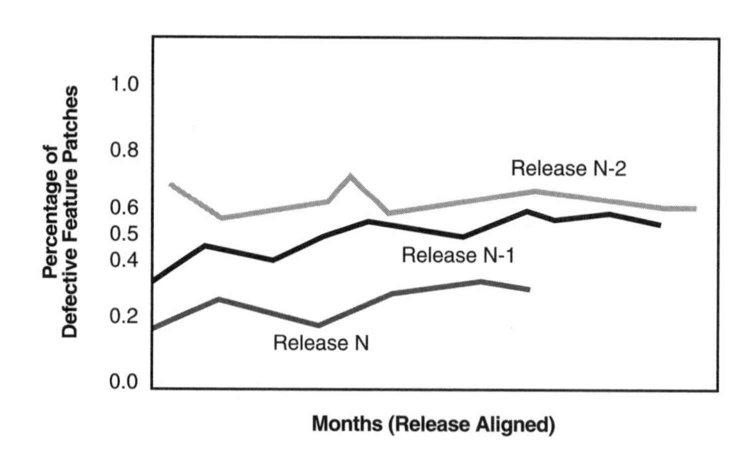

Figure 4.11 Feature patch quality example.

SERVICES MEASUREMENT

There is only one measurement that is applicable to services: service quality.

Service Quality

The service quality measurement is applicable only to services. It can be applied to any type of service provided by a supplier. Some of the quantities measured are:

- Percent conforming installation audits
- Percent successful maintenance visits without callbacks
- Percent successful repairs
- Percent conforming call center calls resolved within agreed-upon time
- Percent support service transactions without a defect

The purpose of this metric is to provide quality measurement information for establishing the evaluation and continuous improvement of the service. The following are examples of the service quality metric:

A. *Installation Example:*

1. Data collected and results

Example Source Data for Installation SQ

	January	February	March	April
# of Nonconforming installation audits	5	1	0	6
Total number of installation audits	100	50	75	80
Service quality metric	95%	98%	100%	92.5%

2. Computation for the month of January:

$$100 \times (1 - 5/100) = 95\%$$

B. *Engineering Example:*

1. Data collected and results

Example Source Data for Engineering SQ

	January	February	March	April
# of engineering corrections	5	2	0	4
Total number of billable engineering staff hours (100s)	10.00	5.00	7.50	3.00
Service quality ratio metric	. 5	.4	0	1.333

2. Computation for the month of January:
   5/10.00 = 50 defects per 100 staff hours

C. *Maintenance Example:*

1. Data collected and results

Example Source Data for Maintenance SQ

	January	February	March	April
# of callbacks	2	0	1	4
# of maintenance visits	30	20	75	120
Quality service conformance metric	93.3%	100%	98.7%	96.7%

2. Computation for the month of January:
   $100 \times (1 - 2/30) = 93.3\%$

D. *Call Center Example:*

1. Data collected and results

Example Source Data for Call Center SQ

	January	February	March	April
# of call resolutions that exceeded the specified time allotment	15	40	10	4
Total # of calls that came into the call center	2000	5000	2750	3000
Quality service conformance metric	99.25%	99.2%	99.6%	99.9%

2. Computations for the month of January:

$$100 \times (1 - 15/2000) = 99.25\%$$

E. *Support Services Example:*

This example references a cable locator service with a defined defect as a cut cable due to incorrect identification.

1. Data collected and results

Example Source Data for Support Services SQ

	January	February	March	April
Cut cables *(# of defects)*	5	2	0	4
# of cables identified *(# of opportunities for defects)*	1000	500	750	300
Quality service conformance metric	99.5%	99.6%	100%	98.7%

2. Computation for the month of January:

$$100 \times (1 - 5/1000) = 99.5\%$$

METRICS SUBMISSION AND ANALYSIS

Obviously, for all of this to work, each party involved in the process, that is, the supplier, their customers, the registrars, and the database administrator, must all perform certain tasks and meet certain responsibilities.[8]

The metrics work group chartered a subgroup to create specifications for the metrics repository system, the registration repository system, and the confidentiality and security aspects of data. The metrics subgroup prepared an elaborate specification for such a system. The next chapter describes these details.

ENDNOTES

1. Malcolm Baldrige National Quality Award, *Criteria for Performance Excellence* (Gaithersburg, MD: Baldrige National Quality Program, 2000).
2. *The European Foundation for Quality Management* (Brussels, Belgium: 1999).

3. GR-929-CORE, *Reliability and Quality Measurement for Telecommunications Systems (RQMS)*, Issue 4 (Morristown, NJ: Telcordia Technologies, 1998).
4. GR-1315-CORE, *In-Process Quality Metrics (IPQM)* (Morristown, NJ: Telcordia Technologies, 1995).
5. GR-1323-CORE, *Supplier Data—Comprehensive Generic Requirements*, Issue 1 (Morristown, NJ: Telcordia Technologies, 1995).
6. V. Basili and D. M. Weiss, "A Methodology for Collecting Valid Software Engineering Data," *IEEE Transactions on Software Engineering SE-10*, no. 6 (November 1984): 728–38.
7. The QuEST Forum, *TL 9000 Quality Management System Measurements Handbook*, Release 3.0 (Milwaukee: ASQ Quality Press, 2001).
8. Galen Aycock, Jean-Normand Drouin, and Thomas Yohe, "TL 9000 Performance Metrics to Drive Improvement," ASQ *Quality Progress* 32, no. 7 (July 1999).

5

TL 9000 Measurements Data Management and Analysis

MEASUREMENTS REPOSITORY SYSTEM

The University of Texas at Dallas (UTD) is responsible for the measurements repository system (MRS). UTD was selected by the QuEST Forum from a pool of very competent contenders. The most sensitive aspect of the MRS is data security. Due to intensive competition between suppliers, they are not comfortable sharing data with each other.

In order to address these concerns, the measurements work group chartered a subteam to define requirements for the MRS. The team prepared a detailed system specification. The main items in the list included:

- MRS specifications describe type of technology, database, data entry, security, confidentiality, data submission, data validation, and so on
- Security aspects include:
  - Key management: access control
  - Measurements data submission: encrypted keys, who issues encrypted keys, and who uses the keys
  - Measurements data storage: data transfer and storage, temporary and permanent storage

The measurements work group posed the following two primary requirements for the MRS. UTD ensured that they satisfied these requirements:

1. Measurement data must remain secure throughout the process, from its origin through analysis and posting
2. No single person should be able to compromise the security, confidentiality, or reliability of submitted measurement data

UTD complied with these requirements effectively. The data security has three important elements, which are:

1. Confidentiality of source and data
2. Integrity of data
3. Availability of processed data

Confidentiality of Source and Data

The QuEST Forum appointed a subteam to define requirements of the MRS. Confidentiality of source and data was the single most important concern of suppliers. UTD worked with the subteam to alleviate such concerns. UTD incorporated two key design elements, which are:

1. Separation of the administrative system from the data system
2. Encryption

The administrative system, called the registration repository system (RRS), is managed by ASQ. When suppliers decide to apply for TL 9000 registration, they notify ASQ of their intention. ASQ requires suppliers to define the scope of their registration (hardware, software, service, or any combination) and product categories. ASQ then assigns reporting IDs for each product category. The supplier submits the data to UTD using the reporting IDs. UTD does not have knowledge of who is submitting the data, nor does ASQ have knowledge of the data content. This is the essence of the double-blind system. Later in the process, ASQ issues a Data Submission Confirmation to signify the supplier's successful submission of data to UTD.

The TL 9000 measurements database (MRS) is managed by UTD. The supplier encrypts the data prior to submission, using the data submission software provided by UTD. The supplier chooses the encryption key, which is 128 bits long. UTD receives the encrypted data from the supplier, but they do not know the owner of the data because the data are transmitted from the incoming server directly into the repository computer. UTD checks for a valid reporting ID with the data. They also perform a sanity check on the data. If the data are valid, they provide a qualitative data confirmation report to ASQ. The MRS and RRS are tied together through the reporting ID in the double blind system.

UTD developed a secure process to ensure that the database is backed up and recoverable in the event of natural disasters. UTD also instituted methods to protect the database from infection by software viruses.

Encryption Process

The encryption process requires that the submitter choose an encryption key that is 128 bits long (16 ASCII characters). The submitter checks and encrypts their data. The submitter is then required to encrypt their own encryption key using public/private encryption of 1024 bits. The encrypted encryption key, along with the encrypted data, is then sent to UTD using one of the following options:

- Anonymous submittal of encrypted data:
  - Transmission of data via internet
  - Mailing of data via courier (FedEx, and so on)
- Non-anonymous submittals of encrypted data:
  - Direct transmission of data via the internet
  - Direct mailing of data via courier

At the other end at UTD, the reporting ID is the only identifying element. UTD does not have knowledge of who owns which reporting ID. Only data with verifiable reporting IDs are accepted by the database program. Also, only encrypted verifiable data from the supplier enter the database. No other data will be accepted by the program. The input data are not seen by anyone at ASQ or UTD.

In summary, the encryption process ensures the following:

- Data is always encrypted
- Unencrypted input data are never shown or seen

Data Security

UTD has made elaborate arrangements to house the MRS system. They have installed stand-alone computers in a locked room. Both encrypted data and computers reside in a locked room. The computer room is under constant video surveillance. The videotapes are regularly viewed for illegal entry. The entry logs are verified twice a week. Nobody has unauthorized access to the computers. All software is strictly controlled.

The system consists of triple-redundant computers. The encrypted data are backed up. The storage computers receiving encrypted data are not connected to any network, so hackers can never access the system. The computer room power is protected by double backup power supplies. Also, the staff handling data is specially trained on the MRS process.

The MRS system complies with BS 7799,[1] the international standard on information security. BS 7799, first published in February 1995, is a comprehensive set of controls comprising best practices in information security. BS 7799 is intended to serve as a single reference point for identifying a range of controls needed for most situations where information systems are used in industry and commerce, and can be used by large, medium, and small organizations. The current document was significantly revised and improved in May 1999.

As part of the BS 7799 assessment, UTD's MRS system was audited by the British Standards Institution (BSI) assessors. BS 7799 covers all aspects of data security through 130 auditable controls. UTD achieved BS 7799 certification in January 2000. The metrics repository system at UTD is the first BS 7799–certified secure system in the United States. They will be audited semiannually by BSI. ASQ also obtained BS 7799 certification for the RRS system in 2001.

METRICS DATA STORAGE

Encrypted supplier data submitted using the Internet is instantaneously transferred to an intermediate computer not accessible via the Internet. The data are then physically moved to temporary storage at the stand-alone, highly secure computer system without decryption. Data submitted via anonymous FedEx shipments are stored only on the stand-alone computer system without decryption.

Supplier data are stored in the secure database when authorized. Storing data in the secure database can only be activated by two persons, for example, two keys must be entered by two key holders for authorization. A two-person password is required for database access to ensure that no single person can compromise (decrypt) the data.

The supplier encryption keys are recovered first to decrypt the data. The data are then encrypted using the UTD key ([N, K] scheme) and stored. This is a password security scheme where N is the total number of passwords and it takes any K passwords to achieve access. Each password is 16 characters long.

Table 5.1 Sample data confirmation report.

MetricID	NPR	CustomerBase	Total
NPRS	OK	OK	OK
NPRA	OK	OK	OK
Np1	Out-of-range	OK	Missing
Np2	OK	OK	OK
Np3	OK	OK	OK
Np4	Not applicable	Not applicable	Not applicable

UTD sends a data confirmation report to a supplier via ASQ. The report does not contain any data. A sample data confirmation report is shown in Table 5.1 for the number of problem reports (NPR) metric. *MetricID* is a keyword. *NPR* is the metric name (number of problem reports). *CustomerBase* is a keyword referring to the number of customers who are QuEST Forum members. *Total* represents all customers. *NPRS* is the normalization factor. *NPRA* is the annualization factor. *Np1*, *Np2*, *Np3*, and *Np4* refer to the number of critical, major, minor, and service problems respectively. Each item in a column represents a value for one month. The software performs a sanity check on the values and determines if the data points are within the range (*OK*); *Missing; Out-of-range;* or *Not applicable.* The input software identifies range values before the data are sent in.

METRICS DATA FLOW

The metrics system involves data collection, analysis, and reporting activities. Each of these activities consists of interaction between the following five entities:

- Supplier
- Customer
- University of Texas at Dallas (UTD)
- American Society for Quality (ASQ)
- Registrar

The double-blind system was set up to insure that the data are secure and not available to competitors and other unauthorized persons. For the five entities to work together efficiently in the double-blind system, it requires extensive communication and passage of data and information.

The five entities must interact with each other at various times. Figure 5.1 shows the overall system, and Figures 5.2 through 5.7 show the paired and triplet interactions.

There are 18 interactions shown in Figure 5.1. Table 5.2 lists these interactions. We will describe the paired and triplet interactions in the next sections of this chapter. Numbers shown in parentheses in those sections indicate the part of the process covered.

Table 5.2 Interactions during metrics data flow.

Interaction	From	To
1. Data submission software package	UTD ASQ	ASQ Supplier
2. ASQ–UTD communications package	UTD	ASQ
3. Send secret ASQ key	ASQ	UTD
4. Add/invalidate reporting ID	ASQ	UTD
5. Registration profile	Supplier	ASQ
6. Reporting and registration IDs and the password	ASQ	Supplier
7. Secret supplier key for reporting ID	Supplier	UTD
8. Performance results	Supplier Customer	Customer Supplier
9. Metrics submission by reporting ID	Supplier	UTD
10. Data confirmation reports	UTD ASQ	ASQ Supplier
11. Revised metrics submission by reporting ID	Supplier	UTD
12. Clarifications	UTD ASQ Supplier ASQ	ASQ Supplier ASQ UTD
13. Summary data confirmation report	ASQ Supplier	Supplier Registrar
14. Registration certificate	Registrar Supplier	Supplier Customer
15. Registrations: new, amended, lost	Registrar	ASQ
16. Quality report to web (QuEST Forum members may view the report)	UTD ASQ QuEST Forum Web site	ASQ QuEST Forum Web site Supplier members
17. Annual report	ASQ	QuEST Forum Web site
18. Registry	ASQ	QuEST Forum Web site

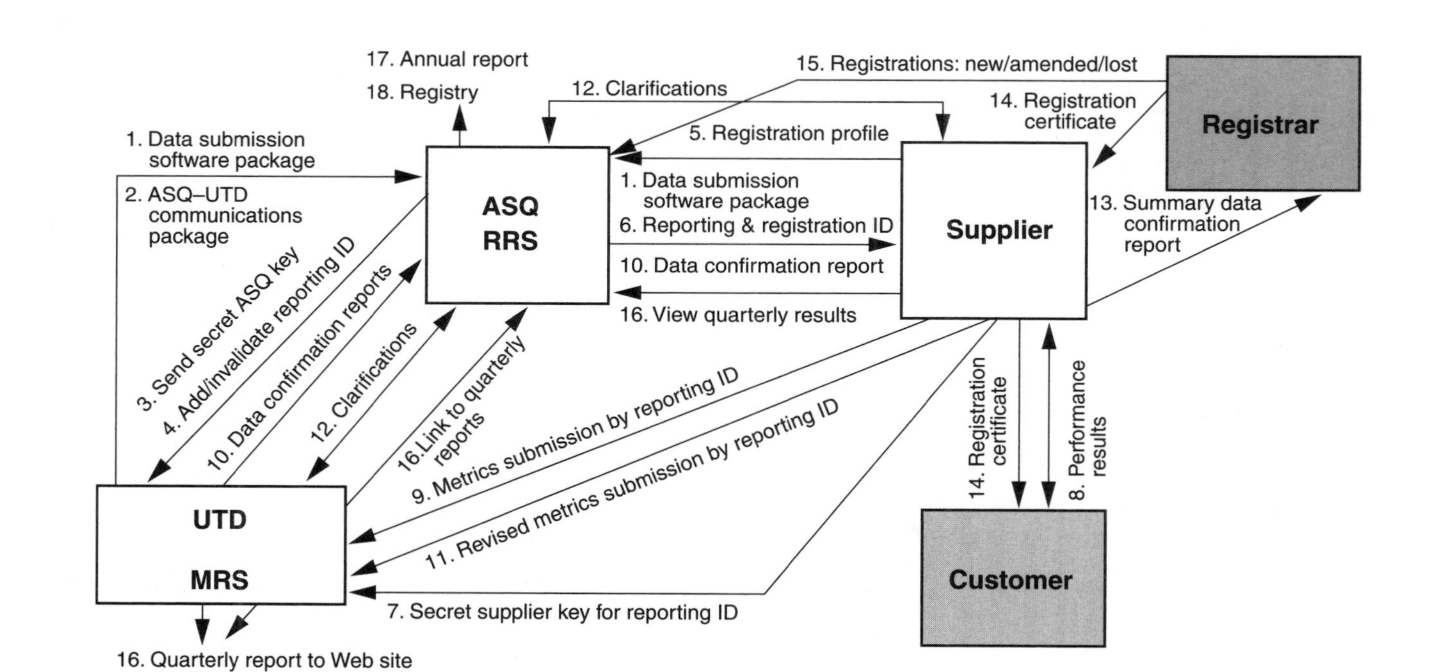

Figure 5.1 Interactions in the metrics implementation process.

Interaction between Supplier and Customer

The interaction between customer and supplier is critical to the metrics system. There are many metrics that require customers to provide the data to the suppliers. Also, TL 9000 requirements mandate stronger customer and supplier relationships.

For example, computation of number of problem reports (NPR) requires that only customer-originated problem reports be counted. So, it is imperative that customers have processes in place to capture problem reports, and then provide necessary data to suppliers to generate the measurement.

The interaction starts when a customer inquires, requests, or contracts with suppliers for TL 9000 registration. The supplier notifies the customer of its plans for TL 9000 registration and requests data to complete the measurement calculations. The supplier and customer together need to establish a process to receive and validate the data on a regular basis (8). The supplier validates the data and requests clarifications when necessary.

The supplier completes the required TL 9000 measurement computations. When the supplier receives certification, it notifies the customer (14). The supplier may meet with customers to provide customer-specific reports with industry norms received from UTD. The supplier and customer also may jointly establish improvement teams. The interaction between customer and supplier is depicted in Figure 5.2.

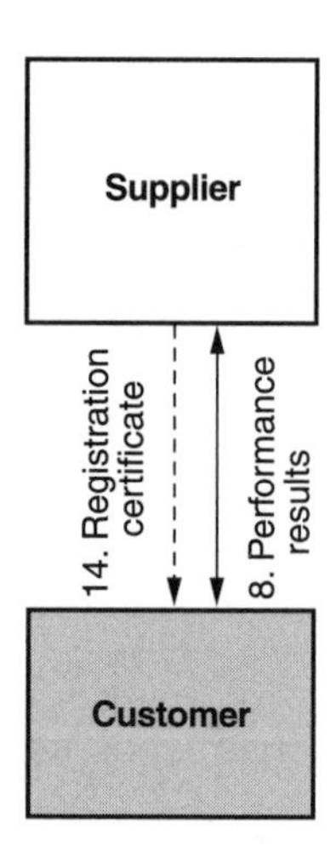

The dashed arrow indicates the flow of information on request.

Figure 5.2 Interaction between customer and supplier.

Interaction between Supplier and ASQ

The relationship between the supplier and ASQ begins when the supplier establishes an account with ASQ and provides a registration profile (5). The information provided by the supplier includes:

- Applicable metrics
- Determination of which will be TL 9000 or RQMS alternative
- Scope of the customer base for metrics

ASQ provides the supplier with the following:

- Access to registration profile with registration ID and password (6)
- Reporting ID for each product category (6)
- Data submission software (DSS) package (1)
- Access to quality results—for QuEST members only (16)

Also, ASQ interfaces with the supplier to obtain clarification (12) of any data concerns. ASQ reroutes the UTD data confirmation report for each reporting ID (10) and mails a summary data confirmation report for each registration. A pictorial representation of this interaction is shown below in Figure 5.3.

Interaction between Supplier and UTD

The supplier uses data submission software (DSS) (1) to:

- Transmit secret supplier key to UTD (7)
- Validate metrics data
- Encrypt metrics data and transmit them to UTD (9)

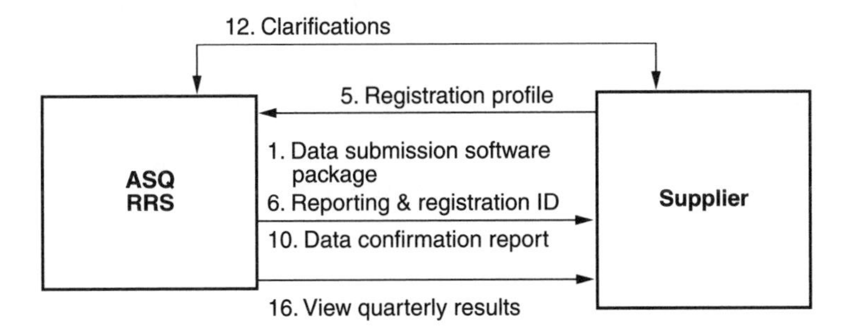

Figure 5.3 Interaction between supplier and ASQ.

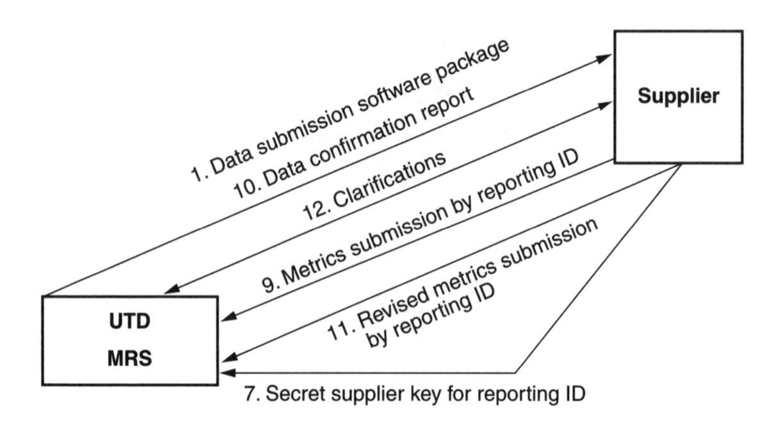

Figure 5.4 Interaction between supplier and UTD.

UTD in turn does the following:

- Sends a confirmation report back to the supplier via ASQ if the transmission is successful (10)
- Validates metrics data
- Identifies missing or out-of-range data
- Asks the supplier (through ASQ) to send revised data or other clarifications if needed (12)

Also, the supplier provides revised measurements to UTD in case of an error or transmission failure (11). If the transmission fails, the supplier works through ASQ to resolve the issues (12). When data has been successfully accepted by UTD, a data confirmation report is sent to the supplier (10).

These interactions are summarized in Figure 5.4.

Interaction between Supplier, ASQ, and UTD When Data Transmission Fails

If the transmission fails, the supplier works through ASQ to resolve the issues (12). UTD in turn does the following:

- Validates measurements data
- Identifies missing or out-of-range data
- If needed, asks supplier (through ASQ) to send revised data (12)

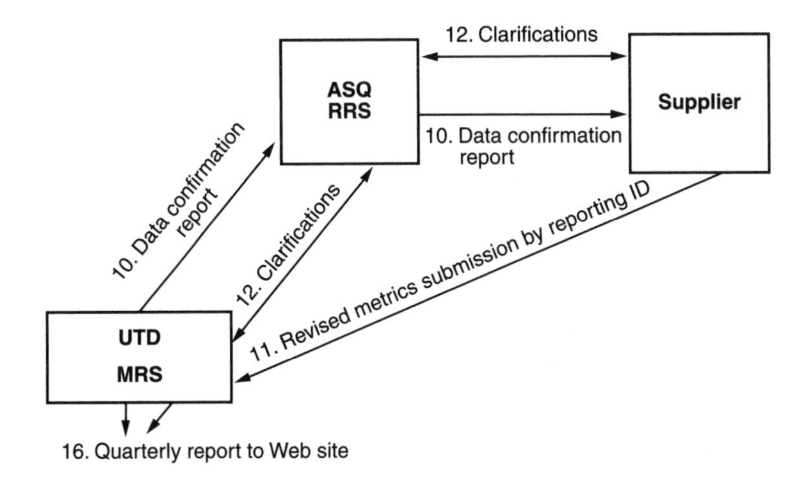

Figure 5.5 Interaction between supplier, ASQ, and UTD when transmission fails.

The supplier resubmits measurements data to UTD (11). If data errors continue, UTD consults with ASQ. When data have successfully been accepted by UTD, a data confirmation report is sent to the supplier (10). The three-way interactions are shown in Figure 5.5.

Interaction between Supplier, Registrar, and ASQ

The interaction between the supplier, registrar, and ASQ is quite extensive. The supplier sends ASQ the registration profile (5). After receiving the registration ID (6) from ASQ, the supplier submits to assessment by the registrar. The registrar assesses the supplier's compliance to:

- TL 9000 requirements
- TL 9000 measurements

The registrar also assesses the supplier's data validity and integrity processes. They verify that measurements are reported for old and new products and review the summary data confirmation report (13) as a quality record.

The registrar writes nonconformances if necessary and ensures that the supplier provides an action plan to address the deficiencies. The supplier is also required to demonstrate to the auditor that the nonconformances against their system are resolved in a timely manner. After a satisfactory assessment, the registrar grants registration (14) to the supplier and notifies ASQ of registration (15). The interactions between these three entities are captured in Figure 5.6.

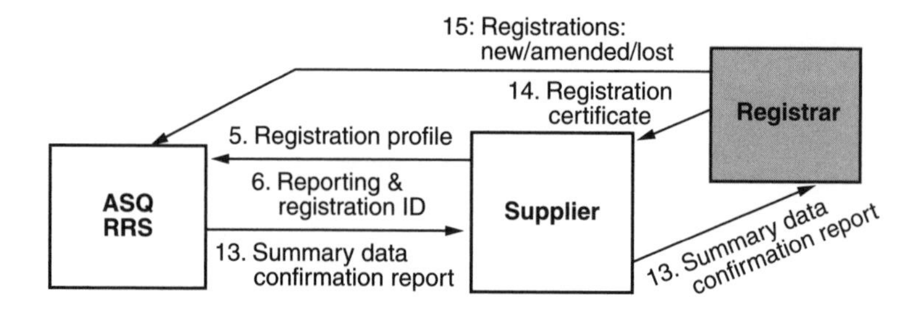

Figure 5.6 Interaction between supplier, ASQ, and registrar.

Interaction between ASQ and UTD

The interaction between ASQ and UTD is very crucial for the MRS system. UTD has defined the communication process between UTD and ASQ. UTD provides the communication package to ASQ (2). ASQ uses the communication package to provide the following to UTD:

- Secret ASQ key (3)
- Valid reporting ID (4)

In addition, ASQ receives the data submission package from UTD (1), which they make available for suppliers (with valid key) through the Web site. UTD and ASQ work jointly to resolve any chronic issues with supplier data (12). UTD also compiles quarterly reports and provides ASQ with a web link (16). ASQ provides gatekeeper functions to access quarterly data at the Web site (16). Later, UTD sends ASQ a data confirmation report (10). See Figure 5.7.

ASQ's Additional Responsibilities

In addition to the interaction with other entities, ASQ also is responsible for publishing the annual report (17) and the registry (18) on the QuEST Forum Web site.

THE GOALS AND EXPECTATIONS OF TL 9000 MEASUREMENTS

There are five major goals defined by the QuEST Forum:

1. To drive continuous improvement in products, services, and processes
2. To develop standard measurements to drive product quality improvement

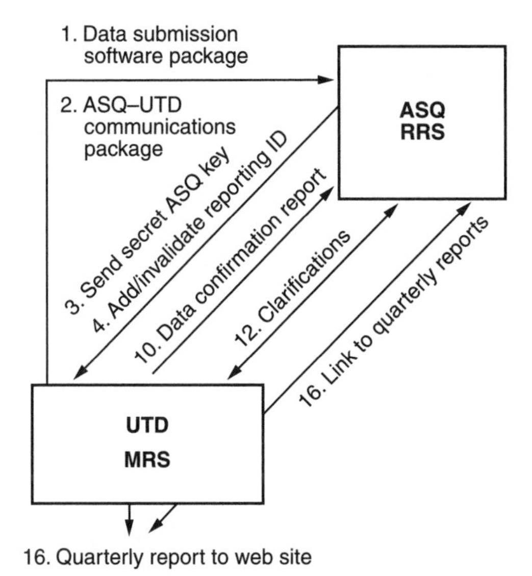

Figure 5.7 Interaction between ASQ and UTD.

3. To implement an industry standard assessment process that would reduce the multiplicity of conflicting programs
4. To foster systems that will protect the integrity of telecommunications products, services, and networks
5. To develop requirements that will help organizations more accurately assess the implementation of their quality management system

There are a number of industry benefits that the QuEST Forum membership expects from the development of TL 9000. These include:

- Improved service to end users
- Enhanced customer–supplier relationships (note: QuEST Forum has been a cooperative effort between customers and suppliers)
- Uniform performance- and cost-based measurements to use as benchmarks for improving product and service quality
- Enhanced supply chain management, including second- and third-tier suppliers
- The creation of a platform for industry improvement initiatives

Specific expectations for Forum member products and services include:

- Cycle-time reduction
- Improved on-time delivery
- Reduced lifecycle costs
- Increased profitability and market share
- Superior products
- Defect reduction

The Use of Measurements to Improve Product and Service Quality

The unique difference of TL 9000 from current ISO 9000 practice is the use of measurements as a benchmarking tool. Measurements are used to determine the quality level of telecommunications products and services. Examples of key measurements are product return rate, number of problem reports, and software update quality. It is expected that extending ISO 9000 in this direction will result in an industry drive to improve overall quality. Certainly when suppliers know where their products stand relative to their competitors and their customers know the quality level of "best in class," there will be a drive by all to improve.

The QuEST Forum envisions multiple uses for the measurements. For example, they can be used to enhance customer–supplier communication that will be used to prioritize and solve the most cost-effective problems. Measurements can also be used to bridge the gap between quality issues and business results. And, they can be used to quantify the customers' views of their suppliers' quality. Some service providers have indicated that they will use the measurements to develop report cards for their suppliers.

The QuEST Forum has retained the University of Texas at Dallas (UTD) to gather the data and develop statistics for each product class. The statistics will identify the mean (or median) standard deviation (or range) and "best in class." The data will be published on the QuEST Forum Web site (www.questforum.org).

A sample plot showing the three measurements for the on-time delivery metric is shown in Figure 5.8.

The key to confidentiality is that no one knows where a particular supplier's data falls on this chart. Each supplier alone knows where they stand on the plot.

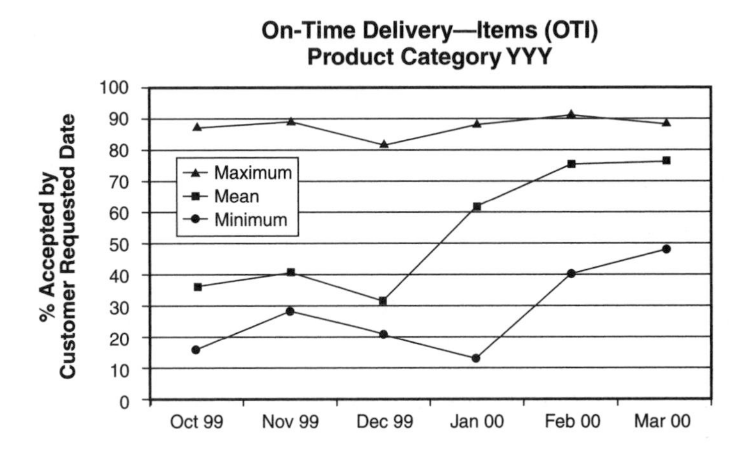

Figure 5.8 Sample on-time delivery measurement.

Also, output from the measurements database consists of monthly or quarterly statistical summary reports derived from the TL 9000 measurements repository for each measurement by product category as posted on the Quest Forum Web site. These reports are posted on the Web site and distributed in hard-copy form by ASQ as appropriate. However, the statistics for a product category are computed and posted only if it can be assured that the statistical results will not in any way compromise supplier anonymity. There must be at least five sets of data from three companies before computations can be made.

UTD was selected because of its long-standing role in telecommunications engineering education. It's expected that UTD will provide insight into benchmarking industry quality problems, identifying quality trends, and identifying possible solutions to quality problems.

Measurements Computation

A major issue during the development of these measurements is the identification of "normalization factors," which will allow comparison of similar products with different characteristics. For example, for circuit switching, the normalization factor for return rate is returns/10,000 terminations/year. Thus products with a large number of terminations per circuit pack are measured fairly against those with fewer terminations per circuit pack.

Another issue is a result of a requirement of some service providers that the suppliers provide "Reliability and Quality Measurements for

Telecommunications Systems" (RQMS) measurements defined in the Telcordia Technologies document GR-929-CORE.[2] In this case, the supplier may provide the RQMS measurements in place of the TL 9000 measurements.

Other measurements and indicators are under consideration, but have not been approved. Indicators are a separate category of measures that are used to "flag potential cost, schedule, productivity, and quality issues."[3]

Product and Service Categories

Along with the definition of metrics, it was an enormous task to identify products for which the metrics would be applicable. A subteam was chartered to create a list of such products. The team spent quite some time doing research. Their deliverable was the product category table.[4,7] The table includes all the products in the telecommunications industry that fall under the metric umbrella. With the frequent introduction of new products it will be a true challenge to keep the product category table current.

In the beginning, the product category table was being updated almost on a monthly basis. Finally, the Forum decided to maintain the latest version of the table on their Web site (www.questforum.org). The list continues to increase as more and more products are introduced.

The following is a partial list of product and service categories defined for metrics implementation (see Appendix A in the *Measurements Handbook*):[7]

- *Switching*: Equipment for the physical or virtual interconnection of communications channels in response to a signaling system. The switching category is broadly defined to include packet- or circuit-switched architectures.

- *Signaling*: Equipment for the provisioning of signaling, that is, "[network] states [that are] applied to operate and control the component groups of a telecommunications circuit to cause it to perform its intended function. . . . There are five basic categories of signals: . . . supervisory, information, address, control, and alerting . . . includes [only] signaling products that function within the telecommunications network. . . ."

- *Transmission*: Equipment for the connection of the switched and interoffice networks with individual customers. An integral part of the distribution network is the loop, which connects the customer to the central office (CO), thus providing access to the interoffice network.

- *Operations and maintenance*: Equipment, systems, and services for the management, upkeep, diagnosis, and repair of the communications network.
- *Common systems*: Any of a variety of specialized generic, shared equipment to support network elements. Common systems include power systems and network equipment–building systems (NEBS) that provide space and environmental support for [the] network.[5]
- *Customer premises*: Equipment installed beyond the network demarcation point. Although commonly installed on the subscriber's premises, equipment with essentially identical functionality installed in the service provider's facility may also be classified as customer premises equipment.
- *Services* (installation, engineering, maintenance, repair, call center, and support): Results generated by activities at the interface between the supplier and the customer and by supplier internal activities to meet customer needs. Note that:
  - The supplier or customer may be represented at the interface by personnel or equipment
  - Customer activities at the interface with the supplier may be essential to the service delivery
  - Delivery or use of tangible product may form part of the service delivery
  - A service may be linked with the manufacture and supply of tangible product
- *Components and subassemblies* (includes component suppliers, contract manufacturers, and OEM suppliers): Individual components or assemblies provided for use in telecommunications systems, excluding those already covered by a specific product category in another product family. These items would typically be used by other suppliers and not sold directly to service providers except as replacement parts.

Training

Since the metric requirement is unique to TL 9000, the QuEST Forum arranged training for suppliers, registrars, and trainers. The Forum appointed two training organizations, Excel Partnership and STAT-A-MATRIX. The

Forum worked with the two training organizations and UTD to develop a number of metrics-related courses. Some of the courses are:

- *Metrics Collection and Analysis*—focused on the planning aspect associated with design and implementation of metrics collection and reporting systems
- *QuEST Forum Sanctioned TL 9000 Metrics Data Submission*—focused on how to perform data submissions, how the security system was installed for MRS, and how to use data for internal improvements
- *Metrics Implementation*—designed to help suppliers understand the importance of metrics and help them design a system for TL 9000 metrics collection, analysis, reporting, and submission

The Forum was actively involved during the course development to ensure that a correct and consistent message is given out regarding TL 9000 measurements. The pilot organizations took early versions of the courses as part of the overall pilot program. They provided valuable input to the course developers, which was later used to improve the courses.

THE QUEST FORUM'S GLOBAL APPROACH

The Forum has also developed a cooperative relationship with the EIRUS organization, a telecommunication metrics user group in Europe. EIRUS uses two sets of metrics based on Telcordia Technologies requirements: (1) European In-Process Quality Metrics (E-IPQM), and (2) European Reliability and Quality Measurements for Telecommunications Systems (E-RQMS).[6] Joint working groups have been established to develop alignment of the EIRUS and QuEST Forum measurements.

ENDNOTES

1. BS 7799:1999, *Information Security Management, Part I: Code of Practice for Information Security Management and Part II: Specification for Information Security Management* (London: The British Standards Institute, 1999).
2. GR-929-CORE, *Reliability and Quality Measurements for Telecommunications Systems (RQMS)*, Issue 4 (Morristown, NJ: Telcordia Technologies, 1998).

3. Henry Malec, "TL 9000 Database Repository and Metrics," *The Informed Outlook* 4, no. 6 (June 1999): 4.
4. The QuEST Forum, *TL 9000 Quality System Metrics*, Book Two, Release 2.5 (Milwaukee: ASQ Quality Press, 1999).
5. GR-63-CORE, *Network Equipment-Building System (NEBS) Requirements: Physical Protection*, no. 1 (Morristown, NJ: Telcordia Technologies, 1995).
6. Information concerning the *European Quality Metrics*: E-IPQM and E-RQMS can be obtained from EURSECOM GmbH, Scloss Wolfsbrunnenweg 35, 69118 Heidelberg, Germany, and its Web site www.eurescom.de/.
7. The QuEST Forum, *TL 9000 Quality Management System Measurements Handbook*, Release 3.0 (Milwaukee: ASQ Quality Press, 2001).

6

TL 9000 Case Studies: An Innovative Way of Gathering Feedback Information

INTRODUCTION

In February 1999, the QuEST Forum took an extraordinary step: it launched a pilot program initiative prior to general publication of the TL 9000 handbooks. Eleven telecommunications suppliers and two telecommunications service providers (TSPs) were enlisted to implement TL 9000 within one or more facilities and with the goal of obtaining registration. The pilot program was started before publication of the first editions of *TL 9000 Quality System Requirements*[1] in May 1999 and *TL 9000 Quality System Metrics*[2] in November 1999.

The pilot program had multiple feedback goals. First, it was intended to obtain information concerning implementation issues. The pilot organizations held monthly meetings to provide information to the pilot participants. Secondly, it was a training ground for the accreditation agencies, the registrars, the training organizations, and the data gatherers. Finally, it was a means of recognizing the organizations participating in the program.

Author's note: This chapter is an adaptation of an article written by Jim Mroz that appeared in *The Informed Outlook*, April 2000.* The author gathered the survey data and collaborated with Jim in the development of the article. I want to thank Jim and the publishers, Robin and Jim Gildersleeve, for working with us to develop the material in this chapter.

---

*"TL 9000 Pilot Program Case Studies Enhance Focus," *The Informed Outlook* 5, no. 4 (April 2000): 1, 27–33.

The pilot program is an example of the proactive efforts the QuEST Forum has made to ensure that the TL 9000 system actually raises the baseline, produces benefits for its users, and measures QMS results to accelerate company- and sectorwide improvements. The pilot initiative provided the QuEST Forum with feedback on TL 9000 that was used to make future revisions to the handbooks. The implementation and registration efforts concluded with recognition at a QuEST Forum meeting in Dallas in January 2000. Sixteen organizations from 11 companies were honored. As an example of the extraordinary cooperation of companies in the QuEST Forum, the pilot organizations agreed not to publicize their registrations until the January meeting.

The QuEST Forum's working groups used project planning techniques similar to those used by ISO Technical Committee (TC) 176 when it revised the ISO 9000 series.

However, the QuEST Forum's use of a pilot program during the development of its sector-specific requirements and metrics demonstrated a unique approach to "standards development." This is in contrast to the validation program used by ISO/TC 176 to gain experience in applying the revised ISO 9001. In that program, the feedback to TC 176 was based on a pseudo-implementation where the participants did a gap analysis, estimated cost and time to implement, and answered an extensive questionnaire. But the participants were not required to implement ANSI/ISO/ASQ Q9001-2000.[3]

PILOT PARTICIPANTS

Eleven companies participated in the pilot program. The participants included hardware, software, hardware and software combination, and service organizations. The types of products ranged from wireless transmission equipment, circuit switching equipment, and transmission equipment to fiber optic cable, software products, and installation and engineering services. The organizations were located in the United States, Canada, England, Poland, and the Netherlands. Table 6.1 contains a list of the pilot participants.

CASE STUDY SURVEYS

As part of the effort to gather information for the case studies, the Forum sent surveys to the pilot program participants—both companies that obtained TL 9000 registration and registrars that conducted the

Table 6.1 Pilot participants.

Hardware (8)	Software (3)	Hardware & Software (4)	Service (1)
Adtran	Motorola	Fujitsu Network Communications	Nortel Networks
NEC America	Nortel Networks	Lucent Technologies	
Nortel Networks (2 organizations)	SBC—California	Nortel Networks	
Pirelli Cables & Systems		Marconi Access	
Siecor (2 organizations)			
Tellabs			

assessments—to evaluate implementation and registration experiences. Nine of the 11 TL 9000–registered pilot organizations and five of seven registrars completed and returned the survey. In this chapter we will look at the case study survey results and point out guidance on the general trends detected among the pilot program participants. Keep in mind that continual improvement and customer satisfaction monitoring are requirements of ANSI/ISO/ASQ Q9001-2000.[3] The experiences of the pilot ISO-registered companies with a new layer of QMS requirements may identify opportunities for other organizations, even those not in the telecommunications industry.

The Case Study Survey Results

The QuEST Forum case study survey was intended to assess the pilot program results and gain feedback on the experience of organizations that have implemented TL 9000. The names of the respondents and any references to the organizations discussed below have been intentionally omitted to provide a look at their common experiences and unique outcomes without focusing on a particular company.

The case study survey was distributed electronically to the pilot program participants in January 2000 and later in the year to their registrars. The pilot survey was grouped into 16 categories and consisted of 63 questions. The nine respondents represent 14 of the 16 registrations from the pilot program. The registrar survey questions were grouped into nine categories consisting of 26 questions.

Pilot Survey Questions

1. General Questions (3*) (for example, facility products/services, type of registration, value of registration)
2. Beginning the Journey: Getting Commitments (3)
3. Setting the Scope (3)
4. Selecting the Registrar (2)
5. Assessing the Status (gap analysis) (4)
6. Documenting/Developing Processes (2)
7. Implementing the Process (5)
8. Training (4)
9. Internal Audits (3)
10. Preassessment (4)
11. Corrective Action from the Internal Audit and/or the Preassessment (2)
12. The Certification Audit (5)
13. Reporting TL 9000 Metrics (5)
14. Estimate of Effort (7)
15. Estimate the Value Added by Implementing TL 9000 (4)
16. Review of the Approach (7)

*Numbers in parentheses indicate the number of questions in each category.

Registrar Survey Questions

1. General Questions (2)
2. Beginning the Journey: Top Management Commitment (2)
3. Setting the Scope (3)
4. Preassessment (3)
5. Corrective Action from the Preassessment (2)
6. The Certification Audit (5)
7. TL 9000 Metrics (1)
8. Estimate Your Client's Effort (2)
9. Review of the Approach (6)

Some Facts About the Case Study Survey Respondents

The following information summarizes the characteristics of the case study respondents:

- Organizational size ranged from 35 in one organization to more than 12,350 employees at 15 separate locations in another organization.
- All organizations were registered previously to ANSI/ISO/ASQC Q9001-1994[4] or ANSI/ISO/ASQC Q9002-1994.[5]
- TL 9000 requirements already covered by the QMS ranged from 40 percent to more than 90 percent. CSQP[SM6,7] (Customer Supplier Quality Program—CSQP[SM] is a service mark of Telcordia) and TickIT[10] registrations were cited as factors by those at or above 85 percent.
- ISO 9001 and ISO 9002 registrars were used by 90 percent, but one company switched from its ISO 9001 registrar because the former registrar was not qualified for TL 9000 registration.
- Each organization obtained registrations in one of four TL 9000 categories—hardware (HW), software (SW), services (SV), and combined hardware/software (HW and SW).
- Sixty percent of TL 9000 registration assessments were done as part of a regularly scheduled ISO 9001 and ISO 9002 surveillance audit.
- Fifty percent relied on "in-house" internal auditors, with the others relying on internal auditors from other company facilities, external auditors, or a combination of internal and external auditors.
- Fifty percent had a preassessment done by the registrar.
- Four registrations covered only one facility, while one registration covered more than 12 facilities in four countries.
- Fifty percent plan to extend the scope of their TL 9000 registrations, and 10 percent more may do so.
- Thirty percent integrated TL 9000 with QS-9000[8] and/or ISO 14001[9] to form joint management systems.
- Top management supported all registration efforts, with the source of approval ranging from the President/CEO to the VP/Director of Quality. The registrars agreed that from their perspective, top management support was strong.

As with ISO 9001 and ISO 9002 and other registration efforts, top management support and involvement was critical to TL 9000 implementation and registration. Two examples illustrate this:

1. In one organization, there was apprehension over the possibility of a large resource requirement, which raised questions about the importance of the program—particularly since management did not have customers requesting TL 9000 conformance. Once the benefits were explained, top management was very supportive of the effort.
2. In another organization, corporate-level management gave approval for TL 9000, but the implementation process ran into resistance from middle management in the facility. This required corporate management to strongly state the company's commitment to TL 9000 before the implementation team had complete internal support.

OBTAINING MANAGEMENT COMMITMENT

Questions to address and discuss:

- What are the business benefits your organization will achieve by implementing TL 9000?
- Are customers requesting TL 9000 registration?
- Do major customers even know what TL 9000 is and what its benefits are?
- What are the estimated total resource impact and implementation costs?
- Are there any existing companywide goals that might conflict with or compete for resources with TL 9000 implementation and registration? Can TL 9000 leverage off any of those goals (for example, implementing a new product lifecycle tool)?
- Can TL 9000 be integrated with other initiatives, such as QS-9000,[8] TickIT,[10] or the Carnegie-Mellon Software Engineering Institute Capability Maturity Model?[11]

A risk in having a lack of internal support within some areas of your company is that, when those who are initially resistant to the effort see the results and benefits, their attempts to catch up with the implementation effort could require additional resources. It's important to get the internal support at the start.

Cost versus Value

Registration is often required by a customer, market, and/or industry, but cost and value have been an issue for companies and other organizations deciding whether to implement management systems and seek registration to ISO 9000, QS-9000, ISO 14001, and/or other standards or requirements. Several survey questions asked about the cost and value of TL 9000 registration, and the responses provide useful perspectives.

Costs relating to implementation and registration take several forms, most notably the cost of hiring registrars, consultants, and trainers; the cost of committing internal resources to the implementation and maintenance efforts; and the cost of interruptions to a facility's operations during development, implementation, and registration. The respondents provided information concerning several key cost factors that need to be considered by companies preparing for or considering TL 9000 implementation and registration:

- *Registrar costs*— Sixty percent of respondents had their TL 9000 registration assessment as part of an ISO 9001 or ISO 9002 surveillance audit, thereby reducing registrar travel and other expenses. However, the audits were more costly because TL 9000 requirements and metrics required 2–7 extra auditor days, with most requiring 4–5 added days.

- *Implementation costs*—The approach to implementation varied. The number of full-time employees assigned ranged from zero to four or more with a median of two to three. Meanwhile, the size of the implementation team ranged from 3 to 41 with a median of 10–12. In the survey results, you can clearly see different approaches, from companies relying solely on a small core of full-time implementers to those relying on a team effort with no one committed to the effort full-time. Nevertheless, each approach requires the organization to remove employees from day-to-day operations on at least a part-time basis, which has a short-term affect on productivity.

- *External support*—Although two respondents relied, at least in part, on external support for internal auditing, none of the respondents reported using outside consultants to help with the implementation effort. Clearly, the respondents did not need consultant assistance because their participation in the pilot program and their membership in the QuEST Forum gave them access to industry resources. In addition, respondents had an inside understanding of TL 9000 since all companies participated in the drafting of the TL 9000 handbooks. Contrast this with most companies using ISO 9001 and/or ISO 9002 without actively participating in the revision of the ISO 9000 standards. As a result, many of these companies will rely on consultants to help them update their QMSs to ANSI/ISO/ASQ Q9001-2000. The need for consultants to help with TL 9000 implementation and registration is likely to remain limited for QuEST Forum members as long as the Forum continues to provide support for its members and involves them in TL 9000 development and revision efforts. Examples of QuEST Forum support are the case studies planned for publication in *The Informed Outlook*, including studies of Nortel Networks,[12] Tellabs Operations,[13] and Lucent Technologies[14] that appeared in the May 2000, August 2000, and January 2001 issues.

- *Training*—Although critical to the successful implementation and use of a QMS, training does involve removing employees from day-to-day operations for at least a brief period of time. How much time was involved? Responses ranged from 12–30 minutes to one week for each full-time employee. Most respondents indicated that significant time was devoted to training the implementation team and other key QMS staff (for example, internal auditors), with everyone else receiving limited training. The answers seemed closely connected to whether the company had used Telcordia (formerly Bellcore) quality requirements and/or had been involved in the CSQP℠ process. A company assessing how much training will be required needs to ask itself: "How well acquainted are our employees with requirements similar to those of TL 9000?"
- *Total effort to obtain registration*—Pilot program participants were asked to estimate how much total staff time was required to achieve registration. Because of the newness of TL 9000 and its metrics requirements, only seven in nine provided responses, and their estimates ranged from two staff months to 4.3 staff years, with a median of one staff year. However, the estimates provided do not indicate how much of the effort had been devoted to maintaining the existing QMS and performance measurements systems.

There are a number of values that respondents attributed to TL 9000 implementation and registration. Some translate into direct cost savings and profitability improvements, while others led to improved management of quality and the organization. The following lists identify those values of TL 9000 implementation and registration experienced by pilot organizations that had direct cost savings and resulted in improved management of quality within the organization.

Direct Cost Savings/Profitability Benefits

- Per one organization, replacing CSQP℠ with TL 9000 saves approximately $20,000/year.
- Collecting TL 9000 metrics saves money by eliminating multiple data collection.
- Compliance to TL 9000 enhances customer satisfaction and loyalty and differentiates the organization from its competition. This results in maintaining or improving market share.
- A reduction in registration and internal auditing costs when some organizations combined multiple ISO 9001/2 registrations into a single TL 9000 registration.
- Using metrics and benchmarking significantly improves products.
- An improvement in the new product introduction process.
- Leveraged past quality investments (ISO 9000, CSQP,℠ RQMS, and so on).

Improved Management of Quality and Business Operations

- More consistent product lifecycle management:
  - Better view of customer needs
  - Stronger emphasis on design control
  - Use of a best-practice approach
  - An improved relationship with key internal and external suppliers
  - An improvement in the product traceability processes
  - The use of quality metrics in benchmarking and improvement
  - The use of self-directed teams was enhanced
- Improved business management processes:
  - Implementation resulted in improvement of existing processes

continued

continued

- Measured results (that is, metrics) provided decision-making information to management
- Registration enhanced internal confidence of the staff
- Quality objectives previously believed unachievable were addressed
- Enhanced the discipline required to document the steps taken to improve products
- Enhanced the ability to replicate successes throughout the organization

One respondent made clear that a key value of TL 9000 was in the metrics (measurements): [15] "Our company has been collecting the metrics mandated by TL 9000 in some form or another for years; metrics are an important part of any effective quality system and are used by executives to make business decisions. [Collecting] the TL 9000 metrics will [result in] this review continuing to occur; [in addition], the metrics will now be included in industry benchmarks that will [result in] refined business decisions."

Managing the Implementation Process

While there was variation in how case study survey respondents handled TL 9000 implementation and registration, the following "program" sums up common activities:

- Obtain commitment and support from upper management.
- Identify an executive sponsor for TL 9000 implementation.
- Select a registrar.
- Initially train one or two individuals in the contents of TL 9000. These individuals will then lead the gap analysis described below.
- Obtain input regarding the current status of the QMS.
- Set the scope of the TL 9000 certification. Make sure you understand the definitions of the product categories. Discuss your definition of scope with the registrar.
- Conduct an initial gap analysis of the existing QMS with respect to the TL 9000 requirements and metrics.

- Determine implementation and registration resource requirements and develop a project plan time line.
- Discuss your implementation plans with your registrar.
- Organize the implementation team. As discussed previously, survey respondents took varying approaches to implementing TL 9000, ranging from a small core group that did all the work to a large team with a part-time director. Factors to be considered by a company implementing TL 9000 are how well established the ISO 9001/2–based QMS is, the resources required and complications previously faced when implementing ISO 9001/2, and the gaps in conformance to TL 9000 and the metrics. Depending on these factors, a company may choose to select a small group of full-time implementers or rely on a broad-based team of "part-time" implementers.
- Use training classes to bring the implementation team members and internal auditors up to speed. The purpose of implementation team training is to ensure that the team is consistently implementing the TL 9000 "adders" and the new metrics throughout the organization. Because they have experience with ISO 9000, internal auditors will play a critical early role in gap analysis of the existing QMS against TL 9000's additional requirements and metrics. However, some requirements and metrics may be entirely new to them and the QMS.
- Conduct a detailed gap analysis of the ISO 9001/2–based QMS against all requirements and metrics of TL 9000. The implementation team needs to take the lead in reviewing existing documentation and metrics against TL 9000, with the internal auditors checking the QMS documentation as revisions are made.
- Develop a gap closure plan and ensure that all gaps are closed.
- Prepare the organization for the TL 9000 implementation and registration process. This involves communication with and training of all employees who affect quality so that they understand the changes to the QMS and support the use of TL 9000. Six steps were identified that are recommended for all companies:

    1. Develop a communications plan to introduce TL 9000 and explain implementation
    2. Engage in communication activities, including presentations to senior management about TL 9000
    3. Develop internal training and awareness materials appropriate to different staff levels

4. Conduct in-house training of all management and associates
5. Inform customers of TL 9000 implementation and registration plans
6. Revise the QMS and conduct internal audits of the TL 9000 "adders" and measurements

Benefits of the Gap Analysis and Internal Audits

Results of gap analysis and internal auditing in the pilot organizations indicated that there are several problem areas to be addressed before achieving TL 9000 conformance. These include:

- Inadequate training on revised and new processes
- A lack of documentation (for example, metrics reporting system, disaster recovery, customer satisfaction, lifecycle model, and servicing)
- Establishment of an integrated metrics program, with alignment to RQMS[16] metrics where appropriate
- Significant revision of quality manual
- Creation of procedures related to software and training requirements
- Development of processes to satisfy requirements for customer–supplier communication, collection of customer satisfaction data, and disaster recovery planning
- Implementation of a lifecycle model (LCM)
- Development of a procedure for tracking design changes to specific products in the field

TL 9000 PROCESS DEVELOPMENT AND DOCUMENTATION

Depending on a company's existing QMS—is it registered to CSQP℠ or does it simply conform to ISO 9001?—the number of new procedures that must be developed and documented will vary. However, among the survey respondents, the following processes most frequently were missing or in need of significant updating. The processes causing the most difficulty are listed first:

- TL 9000 metrics and data collection, verification use in improvement programs, management reporting, customer interchange, and reporting to the measurements administrator (*Measurements Handbook*,[17] 3.1 and 3.5.3)*
- Documented quality improvement program (*Requirements Handbook*,[18] adder 8.5.1.C.1)
- Disaster recovery (*Requirements Handbook*, adder 7.1.C.3)
- Advanced quality training (*Requirements Handbook*, adder 6.2.2.C.5)
- Customer satisfaction data (*Requirements Handbook*, adder 8.2.1.C.1)
- Customer communication procedures (*Requirements Handbook*, adder 5.2.C.2)
- Lifecycle model (*Requirements Handbook*, adder 7.1.C.1)
- Risk assessment (*Requirements Handbook*, adder 7.3.1.C.1k [Project Plan] and adder 7.4.1.C.1b [Purchasing])
- Internal course development (*Requirements Handbook*, adder 6.2.2.C.1)
- Control of production and service provision (*Requirements Handbook*, ISO 9001 requirement 7.5.1)
- Customer communication (*Requirements Handbook*, ISO 9001 requirement 7.2.3 and TL 9000 adders 7.2.3.C.1 through 7.2.3.C.4 and 7.2.3.H.1)
- Problem resolution (*Requirements Handbook*, adders 7.3.7.HS.1 and 8.5.2.S.1)
- Support software and tools management (*Requirements Handbook*, adder 7.1.S.3)
- Computer resources for the target computer (*Requirements Handbook*, adder 7.1.S.2)
- Integration planning (*Requirements Handbook*, adder 7.3.1.S.1)

---

*The information in this section originally referred to release 2.5 of the TL 9000 handbook. The information has been converted to the corresponding parts of release 3.0 of the handbook.

The amount and type of additional procedures required may dictate the makeup of the implementation team. In the pilot program, TL 9000 implementation teams' membership varied with the size of the company and the status of its QMS conformance. While a few respondents indicated that the team represented a cross-section of the company, other teams only included one or more of the following types of employees: ISO 9000 internal auditors/coordinators; metrics SMEs (subject matter experts); quality system/quality assurance/software quality personnel; manufacturing/reliability engineers; and production supervisors/operators. One respondent gave each team member a specific organizational role (for example, primary responsibility for training, auditing metrics, and so on), which helped ensure effective implementation. Four aspects of TL 9000 implementation deserve special attention: training, internal auditing, metrics, and the lifecycle model.

Training

Training, both external and internal to companies implementing TL 9000, is critical. The QuEST Forum worked with Excel Partnership and STAT-A-MATRIX to develop sanctioned courses ranging from a general TL 9000 overview course to specific TL 9000 auditing and metrics courses. Ninety percent of the pilot organizations sent employees to one or more of the courses. The other pilot organizations conducted their own internal training.

Internal Auditing

Another critical process was internal auditing, which many survey respondents implemented according to the following general steps:

- Selection and training of the internal auditing team
- Development of internal auditing schedules
- Enhancement of existing (ISO 9001/2) internal audit checklists
- Use of internal audits for gap analysis

Measurements

Some measurements are common to all types of telecommunications companies, while others are specific to hardware, software, or service organizations. Measurements use and the collection, verification, and reporting of data have already been noted as a difficulty for some respondents. This is largely due to the fact that in many cases the measurements required by TL 9000 did not match the measurements they had already been collecting (for example, they were using different data formulas and counting rules). Those respondents that did not have a problem with the measurements acknowledged that they have been using measurements very similar to those in TL 9000 (for example, RQMS) for years. Among those who had difficulty with measurements, the following are examples of what respondents identified as main obstacles:

- Obtaining data for each of the measurements from several different organizations within the company made the process of collecting data very complex. This complexity caused timing issues that could compromise data submission.
- Explaining the measurements to some managers and staff due to a lack of training was often very difficult.
- Developing a secure program to transmit data to UTD was difficult.
- Aligning an already well-established system with the new TL 9000 measurements and then getting everyone used to the new alignment and vocabulary was difficult.

The Lifecycle Model

The ease or difficulty of implementing the lifecycle model (LCM) depended on the existing QMS. Respondents reported that an LCM was already in place in 40 percent of pilot facilities. Of the others, one, with the agreement of its registrar, did not implement the LCM because the scope only covered manufacturing and the organization had no control over design and development and use and replacement. Of those without an existing LCM, the development process was difficult, with the following problems encountered:

- One facility already had some LCM steps outlined, but "end-of-life" planning and migrations of certain systems created complications
- One facility found the entire LCM process difficult and used a stage gate process
- One facility adjusted its design control documentation but still had difficulty with the definition of LCM

PREASSESSMENT AND REGISTRATION RESULTS

Half of the respondents underwent a preassessment audit. The benefits to those companies ranged from:

- Verification that the QMS was in good shape
- To identifying areas of TL 9000 where the registrar differs from the company on interpretations
- To pointing out areas where the company had overlooked something

However, one respondent indicated that there were no benefits from the preassessment because the "registrar did not know enough about the standard."

What were the most significant findings indicated in the preassessments? Preassessments allowed organizations to work with their auditors to understand the interpretations of the adders. In some cases, because the auditors were as new to TL 9000 as the organization, there was a joint effort to formulate an understanding of the intent of the requirements. The following five major findings were reported by respondents:

1. Metrics reporting system was not documented
2. Procedure for documenting disaster recovery (*Requirements Handbook*, adder 7.1.C.3) was not established
3. Procedure for communicating with selected customers (*Requirements Handbook*, adder 5.2.C.2) was not established
4. Inspection and test documentation (*Requirements Handbook*, adder 8.2.4.HV.1) was incomplete, not maintained, or not identified
5. Training content for courses covering hazardous conditions (*Requirements Handbook*, adder 6.2.2.C.6) had not yet been released

When it came to registration results, three respondents stated that they had zero nonconformances during the registration assessments. Only one reported a major nonconformance. Most of the nonconformances had already been identified as obstacles or difficult processes for compliance. The respondents resolved all significant nonconformances.

The registrars felt that the pilot program went well; there were no problems from their standpoint. One registrar stated that "this was as good and as smooth as any pilot program in which [they] have participated." Another stated that "both parties were determined to enable a first-class effort."

RECOMMENDATIONS TO ORGANIZATIONS IMPLEMENTING TL 9000

Eight key suggestions were mentioned:

1. Involve senior management directly in the implementation process, beyond just lending their "support."
2. Achieve a better understanding of resource requirements.
3. Configure the implementation team to include a cross section of the company.
4. Start the metrics program early.
5. Take more time. Some time allocation problems were the result of starting the pilot program while the TL 9000 handbooks were still being drafted. Most pilot organizations would have liked more time to make changes (update the ISO 9001–based QMS) and to take corrective action in response to the audits. Respondents indicated that they would have started training earlier; devoted more time to training, communication/awareness, gap analysis, and implementation; and allowed more time between the preassessment and registration assessment to take corrective actions.
6. Conduct a preassessment.
7. Increase customer awareness of the company's TL 9000 effort.
8. Take the opportunity to combine facilities under a single TL 9000 certificate.

Perhaps the best summary of the pilot program came from a registrar: "TL 9000 is straightforward and should pose no problem if organizations are committed. Organizations should start with the scope definition, product category table, and metrics (measurements) gathering planning before diving into the registration process. Products and metrics (measurements) drive the process and this must be well understood by the organization."

ENDNOTES

1. The QuEST Forum, *TL 9000 Quality System Requirements*, Book One, Release 2.5 (Milwaukee: ASQ Quality Press, 1999).
2. The QuEST Forum, *TL 9000 Quality System Metrics*, Book Two, Release 2.5 (Milwaukee: ASQ Quality Press, 1999).

3. ANSI/ISO/ASQ Q9001-2000, *Quality Management Standards—Requirements*, 3rd ed. (Milwaukee: ASQ Quality Press, 2000).
4. ANSI/ISO/ASQC Q9001-1994, *Quality Systems—Model for Quality Assurance in Design/Development, Production, Installation and Servicing,* 2nd ed. (Milwaukee: ASQC Quality Press, 1994).
5. ANSI/ISO/ASQC Q9002-1994, *Quality Systems—Model for Quality Assurance in Production, Installation and Servicing,* 2nd ed. (Milwaukee: ASQC Quality Press, 1994).
6. GR-1202-CORE, *Generic Requirements for Customer Sensitive Quality Infrastructure*, Issue 1 (Morristown, NJ: Telecordia Technologies, 1995).
7 SR-3535, *Bellcore CSQP^{SM} Program*, Issue 1 (Morristown, NJ: Telecordia Technologies, 1995).
8. Automotive Industry Action Group, *Quality System Requirements, QS-9000*, 3rd ed. (Detroit, MI: Automotive Industry Action Group [AIAG], 1998).
9. ANSI/ISO 14001-1996, *Environmental Management Systems—Specifications with Guidance for Use* (Milwaukee: ASQC Quality Press, 1996).
10. TickIT, *A Guide to Software Quality System Construction and Certification Using ISO 9001:1994*, Issue 4.0 (1998). (DISC TickIT Office at Telephone +44 (0) 20 89967427 or Fax +44 (0) 20 89967429.)
11. Software Engineering Institute, *The Capability Maturity Model: Guidelines for Improving the Software Process* (Pittsburgh, PA: Carnegie Mellon University, 1995).
12. "For Nortel Networks, TL 9000 Means 30 Percent More," *The Informed Outlook* 5, no. 5 (May 2000): 1, 32–38.
13. "Case Study—Tellabs: TL 9000 Registration Seen to Offer Global Competitive Results," *The Informed Outlook* 5, no. 8 (August 2000): 6–11.
14. "Case Study—TL 9000 Registration by Lucent Focuses on Flagship Product," *The Informed Outlook* 6, no. 1 (January 2001): 1, 24–48.
15. With the publication of the 3.0 handbooks, the term "metrics" was replaced by "measurements."
16. GR-929-CORE, *Reliability and Quality Measurements for Telecommunications Systems (RQMS)*, Issue 4 (Morristown, NJ: Telecordia Technologies, 1998).
17. The QuEST Forum, *TL 9000 Quality Management System Measurements Handbook*, Release 3.0 (Milwaukee: ASQ Quality Press, 2001).
18. The QuEST Forum, *TL 9000 Quality Management System Requirements Handbook*, Release 3.0 (Milwaukee: ASQ Quality Press, 2001).

7

Some Final Thoughts

The QuEST Forum was created as an experiment in cooperation among competing suppliers and their customers. The vision was to create a set of standards that could be used to improve the industry. The Forum was innovative in a number of ways, notably the creation of a metrics process that provides measurements for continual improvement.

Quality is an abstract activity carried out by every individual in an organization working at their best. Increased productivity and high customer satisfaction require established processes, effective training, and continual improvement. These principles are at the heart of TL 9000. They make good business sense and are the initial step in the long journey toward business excellence.

Telecommunications organizations worldwide have started to hearken to these principles. The QuEST Forum has experienced a phenomenal growth in the four years of its existence. Starting in January 1998 with 44 members, it had grown to 155 by December 31, 2001. As of that date there were 141 TL 9000 registrations, including organizations from 52 nonmember companies. Sixty-nine (69) organizations from the United States and 47 from the Asia/Pacific region were registered. The remaining registrations were evenly divided between Canada, Europe, and Latin America.

During 2001, the Forum accomplished the following:

- Aligned the *Requirements Handbook* with ANSI/ISO/ASQ Q9001-2000.
- Fine-tuned the current measurements and evaluated the addition of other measures to the *Measurements Handbook.*

- Revised the training materials to align with the changes, especially due to alignment of the *Registration Handbook* with ANSI/ISO/ASQ Q9001-2000.[1]
- Expanded global membership by holding meetings in Berlin and Buenos Aires.
- Expanded the industry membership to include the cable industry, the new service providers, and the new suppliers.
- Extended TL 9000 throughout the supply chain.
- Developed the Business Excellence Acceleration Model (BEAM) aimed at continual improvement of processes, products, and services. BEAM is based on business excellence models such as the Baldrige National Quality Program[2] and provides guidance for excellence within the telecommunications industry.
- Organized an annual QuEST Forum Best Practices Symposium and developed a Web site to communicate excellence within the industry.

In this book we have attempted to shed light on the contents of TL 9000 and have shared valuable tips on how to meet many of the requirements. Building a competitive advantage should be based on solid business strategy and a practical and efficient system of processes that yields desired measurable results. The QuEST Forum hopes to provide the nucleus of such a system.

ENDNOTES

1. ANSI/ISO/ASQ Q9001-2000, *Quality Management System—Requirements*, 3rd ed. (Milwaukee: ASQ Quality Press, 2000).
2. Malcolm Baldrige National Quality Award, *Criteria for Performance Excellence* (Gaithersburg, MD: Baldrige National Quality Program, 2002).

Appendix 1

Tables of Adders to ANSI/ISO/ASQC Q9001-1994 and ANSI/ISO/ASQ Q9001-2000

Table A1.1 depicts the original source for each TL 9000 requirement. Requirements that are being introduced by the QuEST Forum and do not come from one of the sources are shown in the "QuEST Forum" column. The table is organized by the ANSI/ISO/ASQC Q9001-1994 requirements and contains the labels for the corresponding adders that appeared in the year 2000 edition of the document. Table A1.2 is similar except that it is organized by the ANSI/ISO/ASQ Q9001-2000 requirements and shows the corresponding TL 9000 adders based on the ANSI/ISO/ASQC Q9001-1994 requirements. Table A1.3 summarizes the adders to ANSI/ISO/ASQ Q9001-2000.

Table A1.1 TL 9000 adders organized by ANSI/ISO/ASQC Q9001-1994 requirements.

TL 9000 Book One		Source Document						
Adders to ANSI/ISO/ASQC Q9001-1994	**Adders to ANSI/ISO/ASQ Q9001-2000**	**GR-1202**	**GR-1252**	**TR-179**	**ISO 12207**	**QuEST Forum**	**ISO 9000-3**	**ISO 9004-2**
4.1.1.C.1	5.4.1.C.1					X		
C-Note A	5.4.2.C.1-Note 1							
4.2.2.C.1	7.1.C.1				5.2.4.2			
4.2.2.S.1	7.1.S.3			(R)-4.6-7				
4.2.3.C.1	5.4.2.C.2	R2-18						
4.2.3.C.2	5.4.2.C.1	R2-19						
4.2.3.C.3	5.4.2.C.3	R2-20						

continued

TL 9000 Book One		Source Document						
Adders to ANSI/ISO/ASQC Q9001-1994	**Adders to ANSI/ISO/ASQ Q9001-2000**	**GR-1202**	**GR-1252**	**TR-179**	**ISO 12207**	**QuEST Forum**	**ISO 9000-3**	**ISO 9004-2**
4.2.3.C.4	7.1.C.3				6.2.6.1			
4.3.2 C-NOTE B	7.2.2 C-NOTE 1						4.3.2	
4.3.2 C-NOTE C	7.2.2 C-NOTE 2				5.1.5.1			
4.4.1 V-NOTE D	DELETED							
4.4.1.C.1	7.3.1.C.2			4.1.3-4				
4.4.2.C.1	7.3.1.C.1				5.2.4.5			
4.4.2 C-NOTE E	7.3.1.C.1-NOTE 1							
4.4.2 C-NOTE F	7.3.1.C.1-NOTE 2							
4.4.2.C.2	7.3.1.C.3			3.7.1				
4.4.2.C.2 C-NOTE G	7.3.1.C.3-NOTE 1							
4.4.2.C.3	7.1.C.4				5.5.6.1			
4.4.2.S.1	7.1.S.1& 7.1.S.1-NOTE 1			(R) 3.4.3-1				
4.4.2.S.2	7.1.S.2			(R) 3.4.3-3				
4.4.2.S.3	7.3.1.S.1				5.3.8.1			
NEW	7.3.1.S.1-NOTE 1							
4.4.2.S.4	7.3.1.S.2				5.5.5.2-3			
4.4.3 C-NOTE H	7.3.1 C.2-NOTE 1							
4.4.4.C.1	7.3.2.C.1	R2-10						
4.4.4.C.2	7.3.2.C.2				5.3.2.1-2			
4.4.4.H.1	7.3.2.H.1		R2-12					
4.4.4.S.1	7.3.2.S.1				5.3.4.1			
4.4.4.S.2	7.3.2.S.2						4.4.4.c	
4.4.5.S.1	7.3.3.S.1						4.4.5	
4.4.5.V.1	7.3.3.V.1							6.2.3
4.4.7 C-NOTE I	DELETED							
4.4.8 HV-NOTE J	DELETED							

continued

TL 9000 Book One		Source Document						
Adders to ANSI/ISO/ASQC Q9001-1994	**Adders to ANSI/ISO/ASQ Q9001-2000**	**GR-1202**	**GR-1252**	**TR-179**	**ISO 12207**	**QuEST Forum**	**ISO 9000-3**	**ISO 9004-2**
4.4.8. C - NOTE K	7.3.6 C- NOTE 1							
4.4.8.H.1	8.2.4.H.1		R2-13					
4.4.8.H.2	8.2.4.H.2		R2-14					
4.4.8 H- Note L	8.2.4.H.2- Note 1							
4.4.8.H.3	8.2.4.H.3		R2-15					
4.4.9.C.1	7.3.7.C.1		R2-8					
4.4.9.C.2	7.3.7.C.2		R2-11					
4.4.9.H.1	DELETED		R2-16					
4.4.9.H.2	7.3.7.H.1					X		
4.4.9 H- NOTE M	Included in 7.3.7.H.1							
4.4.9.V.1	7.5.1.V.2					X		
4.5.1.S.1	4.2.3.C.1					X		
4.6.1.C.1	7.4.1.C.1		R2-20		5.1.1.8; 5.1.3.1			
4.6.1 C- NOTE O	7.4.1.C.1- NOTE 1							
4.8.H.1	7.5.3.H.1					X		
4.8.H.2	7.5.3.H.2					X		
4.8.HS.1	7.1.HS.1				6.2.1.1–6.2.2.1			
4.8.HS.1 HS NOTE P:	7.1.HS.1- NOTE 1:							
4.8.HS.2	7.5.3.HS.1						4.8	
4.9.H.1	DELETED		R2-25					
4.9.HV.1	7.5.2.HV.1		R2-26					
4.9.HV.2	6.2.2.HV.1	R2-37						
4.9.HV.3	DELETED	R2-38						
4.9.S.1	7.5.1.S.3						4.9	
4.9.S.2	7.3.6.S.1			(R) 3.9.1-2,3	5.5.5.3, 5.5.6.2, 6.2.6.1			
4.9.V.1	7.5.1.V.1					X		
4.9.V.2	7.1.V.1				5.2.4.5			
4.10.1 C- NOTE Q	DELETED							
4.10.1.HV.1	8.2.4.HV.1		R2-5					
4.10.1.S.1	8.2.4.S.1			3.7.1,2				

continued

TL 9000 Book One		Source Document						
Adders to ANSI/ISO/ASQC Q9001-1994	**Adders to ANSI/ISO/ASQ Q9001-2000**	**GR-1202**	**GR-1252**	**TR-179**	**ISO 12207**	**QuEST Forum**	**ISO 9000-3**	**ISO 9004-2**
4.10.4.H.1	8.2.4.H.4		R2-28					
4.10.4.H.2	7.5.5.HS.1					X		
4.10.4.H.2 H-NOTE R	7.5.5. HS.1-NOTE 1							
4.10.5.HV.1	8.2.4.HV.2		R2-42					
4.11.2.H.1	7.6.H.1		R2-31					
4.13.2.C.1	8.4.C.1					X		
4.14.1 C-NOTE S	8.5.1 C-NOTE 1							
4.14.2 C-Note T	8.5.2. C-Note 1							
4.14.2 C-Note U	8.5.2. C-Note 2							
4.15.1.C.1	6.4.C.1		R2-38					
4.15.1.C.2	7.5.5.C.1.		R2-39					
4.15.2.S.1	7.5.5.S.1						4.15.2; 4.15.6	
4.15.3.H.1	7.5.5.H.1		R2-40					
4.15.6.S.1	7.5.1.S.2			3.9.4-2				
4.18.C.1	6.2.2.C.1	R2-36						
4.18.C.2	6.2.2.C.2	R2-34						
4.18.C.3	6.2.2.C.3	R2-32, 35						
4.18.C.4	6.2.2.C.4					X		
4.18.C.5	6.2.2.C.5		R2-56					
4.18.C.6	6.2.2.C.6					X		
4.19.C.1	7.5.1.C.1		R2-52					
4.19.C.2	7.5.1.C.2	R2-16						
4.19.C.3	7.2.3.C.1		R2-54					
4.19.C.4	7.2.3.C.2 & 7.2.3.C.2-NOTE 1			4.10.2-7,8				
4.19.C.5	7.2.3.C.3			4.10.2-6				
NEW	7.2.3.C.4					X		
4.19.H.1	7.2.3.H.1					X		
4.19.HS.1	7.5.1.HS.1	R2-17						
4.19.HS.2	7.3.7.HS.1			4.10.2-9				
4.19.HS.3	7.5.1.HS.2				5.3.12, 5.3.12.2, 5.5.5.4-5, 5.5.6.3-5			

continued

continued

TL 9000 Book One		Source Document						
Adders to ANSI/ISO/ASQC Q9001-1994	**Adders to ANSI/ISO/ASQ Q9001-2000**	**GR-1202**	**GR-1252**	**TR-179**	**ISO 12207**	**QuEST Forum**	**ISO 9000-3**	**ISO 9004-2**
4.19.S.1	7.5.1.S.1			3.10.3-1				
4.19.S.2	8.5.2.S.1			4.10.2.5				
4.20.1.C.1	8.2.3.C.1						4.20	
4.21.1.C.1	8.5.1.C.1		R2-57					
4.21.1.C.1 C -NOTE V	8.5.1 C-NOTE 2							
4.21.1.C.2	8.5.1.C.2	R2-26						
4.21.1.C.3	5.5.3.C.1	R2-31						
4.21.2.C.1	5.2.C.1	R2-7						
4.21.2.C.2	5.2.C.2	R2-9, 15,18						
4.21.2 C-Note W	5.2.C.2-Note 1							
4.21.3.C.1	8.2.1.C.1	R2-46						
4.21.3.H.1	8.4.H.1		R2-46					
4.21.3.V.1	8.4.V.1		R2-46					
4.21.4.C.1	7.1.C.2					X		
4.21.4 C-NOTE X	7.1.C.2-NOTE 1							

Table A1.2 TL 9000 adders organized by ANSI/ISO/ASQ Q9001-2000 requirements.

Adders to ANSI/ISO/ASQ Q9001-2000	Adders to ANSI/ISO/ASQC Q9001-1994	GR-1202	GR-1252	TR-179	ISO 12207	QuEST Forum	ISO 9000-3	ISO 9004-2
4.2.3.C.1	4.5.1.S.1					X		
5.2.C.1	4.21.2.C.1	R2-7						
5.2.C.2	4.21.2.C.2	R2-9, 15,18						
5.2.C.2-Note 1	4.21.2 C-Note W							
5.4.1.C.1	4.1.1.C.1					X		
5.4.2.C.1	4.2.3.C.2	R2-19						
5.4.2.C.1-Note 1	4.1.1.C-Note A							
5.4.2.C.2	4.2.3.C.1	R2-18						
5.4.2.C.3	4.2.3.C.3	R2-20						
5.5.3.C.1	4.21.1.C.3	R2-31						

continued

Adders to ANSI/ISO/ASQ Q9001-2000	Adders to ANSI/ISO/ASQC Q9001-1994	GR-1202	GR-1252	TR-179	ISO 12207	QuEST Forum	ISO 9000-3	ISO 9004-2
6.2.2.C.1	4.18.C.1	R2-36						
6.2.2.C.2	4.18.C.2	R2-34						
6.2.2.C.3	4.18.C.3	R2-32, 35						
6.2.2.C.4	4.18.C.4					X		
6.2.2.C.5	4.18.C.5		R2-56					
6.2.2.C.6	4.18.C.6					X		
6.2.2.HV.1	4.9.HV.2	R2-37						
6.4.C.1	4.15.1.C.1		R2-38					
7.1.C.1	4.2.2.C.1				5.2.4.2			
7.1.C.2	4.21.4.C.1					X		
7.1.C.2-NOTE 1	4.21.4 C-NOTE X							
7.1.C.3	4.2.3.C.4				6.2.6.1			
7.1.C.4	4.4.2.C.3				5.5.6.1			
7.1.HS.1	4.8.HS.1				6.2.1.1–6.2.2.1			
7.1.HS.1-NOTE 1:	4.8.HS.1 HS NOTE P:							
7.1.S.1& 7.1.S.1-NOTE 1	4.4.2.S.1			(R) 3.4.3-1				
7.1.S.2	4.4.2.S.2			(R) 3.4.3-3				
7.1.S.3	4.2.2.S.1			(R)-4.6-7				
7.1.V.1	4.9.V.2				5.2.4.5			
7.2.2 C-NOTE 1	4.3.2 C-NOTE B						4.3.2	
7.2.2 C-NOTE 2	4.3.2 C-NOTE C				5.1.5.1			
7.2.3.C.1	4.19.C.3		R2-54					
7.2.3.C.2 & 7.2.3.C.2-NOTE 1	4.19.C.4			4.10.2-7,8				
7.2.3.C.3	4.19.C.5			4.10.2-6				
7.2.3.C.4	NEW					X		
7.2.3.H.1	4.19.H.1					X		
7.3.1.C.1	4.4.2.C.1				5.2.4.5			
7.3.1 C.1-NOTE 1	4.4.2 C-NOTE E							
7.3.1.C.1-NOTE 2	4.4.2 C-NOTE F							

continued

Adders to ANSI/ISO/ASQ Q9001-2000	Adders to ANSI/ISO/ASQC Q9001-1994	GR-1202	GR-1252	TR-179	ISO 12207	QuEST Forum	ISO 9000-3	ISO 9004-2
7.3.1.C.2	4.4.1.C.1			4.1.3-4				
7.3.1.C.2-NOTE 1	4.4.3 C-NOTE H							
7.3.1.C.3	4.4.2.C.2			3.7.1				
7.3.1.C.3-NOTE 1	4.4.2.C.2 C-NOTE G							
7.3.1.S.1	4.4.2.S.3				5.3.8.1			
7.3.1.S.1-NOTE 1	NEW							
7.3.1.S.2	4.4.2.S.4				5.5.5.2-3			
7.3.2.C.1	4.4.4.C.1	R2-10						
7.3.2.C.2	4.4.4.C.2				5.3.2.1-2			
7.3.2.H.1	4.4.4.H.1		R2-12					
7.3.2.S.1	4.4.4.S.1				5.3.4.1			
7.3.2.S.2	4.4.4.S.2						4.4.4.c	
7.3.3.S.1	4.4.5.S.1						4.4.5	
7.3.3.V.1	4.4.5.V.1							6.2.3
7.3.6 C-NOTE 1	4.4.8. C - NOTE K							
7.3.6.S.1	4.9.S.2			(R) 3.9.1-2,3	5.5.5.3, 5.5.6.2, 6.2.6.1			
7.3.7.C.1	4.4.9.C.1		R2-8					
7.3.7.C.2	4.4.9.C.2		R2-11					
7.3.7.HS.1	4.19.HS.2			4.10.2-9				
7.3.7.H.1	4.4.9.H.2 & 4.4.9 H-NOTE M					X		
7.4.1.C.1	4.6.1.C.1		R2-20		5.1.1.8; 5.1.3.1			
7.4.1.C.1-NOTE 1	4.6.1 C-NOTE O							
7.5.1.C.1	4.19.C.1		R2-52					
7.5.1.C.2	4.19.C.2	R2-16						
7.5.1.HS.1	4.19.HS.1	R2-17						
7.5.1.HS.2	4.19.HS.3				5.3.12, 5.3.12.2, 5.5.5.4-5, 5.5.6.3-5			
7.5.1.S.1	4.19.S.1			3.10.3-1				

continued

Adders to ANSI/ISO/ASQ Q9001-2000	Adders to ANSI/ISO/ASQC Q9001-1994	GR-1202	GR-1252	TR-179	ISO 12207	QuEST Forum	ISO 9000-3	ISO 9004-2
7.5.1.S.2	4.15.6.S.1			3.9.4-2				
7.5.1.S.3	4.9.S.1						4.9	
7.5.1.V.1	4.9.V.1					X		
7.5.1.V.2	4.4.9.V.1					X		
7.5.2.HV.1	4.9.HV.1		R2-26					
7.5.3.HS.1	4.8.HS.2						4.8	
7.5.3.H.1	4.8.H.1					X		
7.5.3.H.2	4.8.H.2					X		
7.5.5.C.1.	4.15.1.C.2		R2-39					
7.5.5.H.1	4.15.3.H.1		R2-40					
7.5.5.HS.1	4.10.4.H.2					X		
7.5.5. HS.1-NOTE 1	4.10.4.H.2 H -NOTE R							
7.5.5.S.1	4.15.2.S.1						4.15.2; 4.15.6	
7.6.H.1	4.11.2.H.1		R2-31					
8.2.1.C.1	4.21.3.C.1	R2-46						
8.2.3.C.1	4.20.1.C.1						4.20	
8.2.4.H.1	4.4.8.H.1		R2-13					
8.2.4.H.2	4.4.8.H.2		R2-14					
8.2.4.H.2-Note 1	4.4.8 H-Note L							
8.2.4.H.3	4.4.8.H.3		R2-15					
8.2.4.H.4	4.10.4.H.1		R2-28					
8.2.4.HV.1	4.10.1.HV.1		R2-5					
8.2.4.HV.2	4.10.5.HV.1		R2-42					
8.2.4.S.1	4.10.1.S.1			3.7.1,2				
8.4.C.1	4.13.2.C.1					X		
8.4.H.1	4.21.3.H.1		R2-46					
8.4.V.1	4.21.3.V.1		R2-46					
8.5.1.C.1	4.21.1.C.1		R2-57					
8.5.1 C-NOTE 1	4.14.1 C-NOTE S							
8.5.1 C-NOTE 2	4.21.1.C.1 C -NOTE V							
8.5.1.C.2	4.21.1.C.2	R2-26						
8.5.2. C-Note 1	4.14.2 C-Note T							
8.5.2. C-Note 2	4.14.2 C-Note U							
8.5.2.S.1	4.19.S.2			4.10.2.5				

continued

continued

Adders to ANSI/ISO/ASQ Q9001-2000	Adders to ANSI/ISO/ASQC Q9001-1994	GR-1202	GR-1252	TR-179	ISO 12207	QuEST Forum	ISO 9000-3	ISO 9004-2
DELETED	4.4.1 V-NOTE D							
DELETED	4.4.7 C-NOTE I							
DELETED	4.4.8 HV-NOTE J							
DELETED	4.4.9.H.1		R2-16					
DELETED	4.9.H.1		R2-25					
DELETED	4.9.HV.3	R2-38						
DELETED	4.10.1 C-NOTE Q							

Table A1.3 Summary of adders to ANSI/ISO/ASQ Q9001-2000.

ANSI/ISO/ASQ Q9001-2000 Element	Common	Software	Hardware	Service	Hardware & Software	Hardware & Service	Total
4.1							
4.2	1						**1**
5.1							
5.2	2						**2**
5.3							
5.4	4						**4**
5.5	1						**1**
6.1							
6.2	6					1	**7**
6.3							
6.4	1						**1**
7.1	4	3		1	1		**9**
7.2	4		1				**5**
7.3	7	6	2	1	1		**17**
7.4	1						**1**
7.5	3	4	3	2	4	1	**17**
7.6			1				**1**
8.1							
8.2	2	1	4			2	**9**
8.3							
8.4	1		1	1			**3**
8.5	2	1					**3**
Total	**39**	**15**	**12**	**5**	**6**	**4**	**81**

Appendix 2

TL 9000 Registration and Accreditation

As you pursue TL 9000 registration, you'll find striking similarities between it and other certification efforts your organization may have undertaken—particularly ISO 9000. The TL 9000 registration period extends over three years, during which time the entire quality management system is examined. The scope of the registration may include hardware, software, services, or any combination of the three, and could apply to part or all of an organization. The scope you choose will be clearly identified on your registration certificate. The number of days required for the audit is based on established guidelines, and any nonconformances must be resolved before you receive your certificate. Trained auditors with expertise in the telecommunications industry will perform the assessment of your organization, and registrars and accreditation bodies must follow criteria set forth by the QuEST Forum.

Further, the QuEST Forum recognizes that there are many organizations with well-established quality management systems in place. Organizations with mature quality management systems may choose to pursue the alternative method for registration to TL 9000 that leverages the organization's internal audit system.

GETTING STARTED

To begin your journey, log on to www.questforum.org, the QuEST Forum Web site. The TL 9000 Users Manual describes the steps you must take to get your effort under way. The Web site is a terrific source of information. Useful Web pages include a general overview of TL 9000, registration instructions and templates, a list of QuEST Forum member companies, a list of TL 9000–qualified registrars, and links to reference materials and training information.

Select your registrar early in the implementation process. Choose a registrar based on their ability to meet your business needs and to provide you with qualified auditors. Discuss your registration scope and approach to TL 9000 with your registrar to ensure that you and your registrar are in agreement in order to prevent surprises during the initial assessment.

Consider the following checklist for TL 9000 audit readiness:

- Registrar/auditors selected and engaged
- TL 9000 requirements effectively implemented
- Metrics data submitted and confirmation report received
- At least one management review performed
- At least one internal audit cycle completed

The Initial Audit

The audit is intended to verify that the TL 9000 requirements have been effectively implemented. In addition to meeting the requirements of ANSI/ISO/ASQC Q10011, parts 1, 2, and 3[4,5,6] (ISO 19011 when approved), the auditors will be experienced ISO 9000 auditors and will have attended and successfully completed QuEST Forum sanctioned training to ensure consistency of interpretation of the TL 9000 requirements. Additionally, at least one member of the audit team must have experience in the telecommunications industry.

The number of days required for the assessment will depend on the number of employees in the organization. An audit-days table based on ISO Guide 62 is located on the QuEST Forum Web site. The table lists the minimum number of on-site days required for initial and ongoing registration and surveillance audits. If the number of audit-days proposed by your registrar exceeds one half of a day less than what is required by the audit-days table, your registrar must submit a deviation request to the Registrar Accreditation Board within five days of providing you with a quote. The accreditation body will provide a written response within 10 days.

When your audit takes place, the auditors will identify opportunities for improvement along with nonconformanees, and document them in a report that they will provide to you within 45 days of the audit. Any major or systemic nonconformances must be resolved prior to your certificate being issued. Your registrar is required to have a process in place to settle any disputes you may have regarding matters of interpretation and nonconformance.

If you have chosen to utilize the services of a consultant for your implementation efforts, be aware that although they may be present during the assessment, their role will be strictly limited to that of an observer.

Migration to TL 9000

If your organization maintains a current registration to an existing quality management system standard, you may follow a migration path to achieve TL 9000 registration. The QuEST Forum currently recognizes ANSI/ISO/ASQ Q9001:2000[1] and QS-9000[2] quality management system registrations. Other registrations such as TickIT[3] will be recognized as the need arises. If your organization is currently registered to QS-9000, for example, your registrar will skip all of the ISO 9001 and customer satisfaction requirements in the TL 9000 set of requirements on its *initial* assessment of your quality management system. Subsequent assessments will incorporate all of the TL 9000 requirements. If, however, your organization is not currently registered to any of the above quality management system standards, *all* of the TL 9000 requirements will be audited during the initial assessment.

Publicizing Your Registration

Once you have achieved TL 9000 registration, you may publicize that fact. Ensure that your advertising states that your quality management system has been registered to TL 9000 and does not in any way suggest that your products have been registered. You must take the scope of your registration into account when deciding how and where to advertise so that you do not imply that nonregistered organizations within your organization are included within the scope of registration. Although you may not use the TL 9000 certification mark on any product or packaging labeling, you may use it on stationery and promotional material. Any misuse of the certification mark is grounds for withdrawal of your certification.

Your Registrar

Your TL 9000–approved registrar is required to meet stringent criteria to ensure quality and consistency in the registration process. All approved registrars must follow the requirements established by the QuEST Forum on how to perform TL 9000 registration audits, and must comply with the "Code of Practice for TL 9000 Registrars" which outlines ground rules for the registration process.

The approved registrars are required to have in-house telecommunications and TL 9000 expertise to better support your efforts. Each registrar must have at least one member of their governing board or council of experts with telecommunications experience who has successfully passed the QuEST Forum sanctioned TL 9000 training.

As part of the registrar qualification process, a QuEST Forum–recognized accreditation body witnesses each registrar during one of its first six TL 9000 audits. Prior to approval, registrars may not use the TL 9000 notation on registration certificates. After approval, registrars will receive a certificate or some other form of notification from the accreditation body. All recognized accredited registrars are listed on the QuEST Forum Web site. Accreditation bodies are subject to similar qualification criteria and must meet the requirements established by the QuEST Forum.

Alternative Method for Maintaining Certification

If you have a mature quality management system, you may opt for the alternative method for TL 9000 registration. The purpose of the alternative method is to eliminate audit duplication and to ensure that your third-party audits are effective by allowing your registrar to utilize the results of your internal audit system as a basis for their assessments. This approach empowers your organization by shifting the responsibility for verifying the effectiveness of your quality management system from the registrar to your organization. The extent to which internal audits may be used as a substitute for external audits will be different for each organization and depends on various factors, including the maturity of the quality management system and the ability of the registrar to verify internal audit results. Advantages to this approach include reduced cost due to fewer external audit-days on site and less disruption to the organization. Recognize that your registrar will continue to conduct on-site audits, but on a limited basis by sampling both TL 9000 requirements and sites.

If you desire to pursue the alternative method, be aware of stringent qualification criteria. Your TL 9000–compliant quality management system must have been in place for at least three years. You must operate an effective internal audit system that adheres to ISO 10011[4,5,6] (ISO 19011 when approved) and have a robust management review process in which goals and objectives are reviewed and updated, and corrective and preventive action is a routine part of the process. Your management must demonstrate their commitment to quality and to improvements of quality objectives and results over at least a two-year period. You must also be able to demonstrate customer satisfaction.

Your registrar will base their audits on the effectiveness of your quality performance over the last three years by considering earlier surveillance audit results, results from independent customer satisfaction surveys, and customer complaints. If your organization fails to meet the alternative method criteria, your registrar will default to the regular TL 9000 registration process.

Your registrar and accreditation bodies must meet similarly challenging criteria to participate in the alternative method approach. Registrars must demonstrate the maturity of their own business management systems by designing a process to support the alternative method accreditation bodies by their ability to verify this capability. Registrars must be accredited to ISO/IEC Guide 62 for at least three years, and must be able to demonstrate improvement of service quality objectives and results over a period of three years.

CONCLUSION

As you journey towards TL 9000 registration, stay informed of the guidelines and resources on the QuEST Forum Web site. Develop a relationship with your registrar early in the implementation process to build an effective partnership. Expect that your auditors, registrar, and accreditation body have met the criteria established by the QuEST Forum to ensure quality TL 9000 assessments. Follow the TL 9000 migration paths where appropriate to save time and money, and strive for meeting the qualification criteria for the alternative method approach to maintaining TL 9000 registration. Good luck on your TL 9000 endeavors.

ENDNOTES

1. ANSI/ISO/ASQ Q9001-2000, *Quality Management Standards—Requirements*, 3rd ed. (Milwaukee: ASQ Quality Press, 2000).
2. Automotive Industry Action Group, *Quality System Requirements, QS-9000*, 3rd ed. (Detroit, MI: Automotive Industry Action Group [AIAG], 1998).
3. TickIT, *A Guide to Software Quality System Constructin and Certification Using ISO 9001:1994*, no. 4.0 (DISC TickIT office at Telephone +44 (0) 20 89967427 or fax +44 (0) 2089967429).
4. ANSI/ISO/ASQC Q10011-1-1994, *Guidelines for Auditing Quality Systems—Auditing* (Milwaukee: ASQC Quality Press, 1994).
5. ANSI/ISO/ASQC Q10011-2-1994, *Guidelines for Auditing Quality Systems—Qualification Criteria for Quality Systems Auditors* (Milwaukee: ASQC Quality Press, 1994).
6. ANSI/ISO/ASQC Q10011-3-1994, *Guidelines for Auditing Quality Systems—Management of Audit Programs* (Milwaukee: ASQC Quality Press, 1994).

Bibliography

"Alcatel Registers All Products, Services, and Facilities to TL 9000." *Quality System Update* 10, no. 7 (July 2000): 4, 17.

Anderson, Stewart. "TL 9000: Reshaping the Telecommunications Industry." *The Informed Outlook* 4, no. 8 (August 1999): 10–12.

ANSI/ISO/ASQ Q9000-3-1997. *Quality Management and Quality Assurance Standards—Part 3: Guidelines for the Application of ISO 9001:1994 to the Development, Supply, Installation, and Maintenance of Computer Software*. 2nd ed. Milwaukee: American Society for Quality, 1997.

ANSI/ISO/ASQC Q9004-2-1994. *Quality Management and Quality System Elements—Part 2: Guidelines for Services.* Milwaukee: American Society for Quality Control, 1994.

ANSI/ISO/ASQC Q10011-1-1994. *Guidelines for Auditing Quality Systems—Auditing*. Milwaukee: American Society for Quality Control, 1994.

ANSI/ISO/ASQC Q10011-2-1994. *Guidelines for Auditing Quality Systems—Qualification Criteria for Quality Systems Auditors*. Milwaukee: American Society for Quality Control, 1994.

ANSI/ISO/ASQC Q10011-3-1994. *Guidelines for Auditing Quality Systems—Management of Audit Programs*. Milwaukee: American Society for Quality Control, 1994.

Aycock, Galen, Jean-Normand Drouin, and Thomas Yohe. "TL 9000 Performance Metrics to Drive Improvement." ASQ *Quality Progress* 32, no. 7 (July 1999): 41–45.

Caplan, Frank. *The Quality System.* 2nd ed. Radnor, PA: Chilton Book Company, 1990.

Evans, Michael W., and John J. Marciniak. *Software Quality Assurance & Management*. New York: John Wiley & Sons, 1987.

"For Nortel Networks, TL 9000 Means 30 Percent More." *The Informed Outlook* 5, no. 5 (May 2000): 1, 32–38.

Gildersleeve, James P. "TL 9000 Quality System Requirements Rooted in ISO 9001." ASQ *Quality Progress* 32, no. 6 (June 1999): 67–68.

GR-1202-CORE. *Generic Requirements for Customer Sensitive Quality Infrastructure*. No. 1. Morristown, NJ: Telecordia Technologies, 1995.

Hutchison, Eugene, and Sandford Liebesman. "Telecom Industry Aims to Save Billions with ISO 9000–Based Standard." *ISO 9000 News* 7, no. 6 (November/December 1998): 1, 20–24.

Jarvis, Alka S. "Improvements Process Based on Metrics." Third Annual Software Quality Conference. Milpitas, CA, May 1993.

———. "Applying Software Quality." *The Seventh International Software Quality Week*. San Francisco, CA, May 1994.

———. "Exploring the Needs of A Developer and A Tester." *Quality Conference*. Santa Clara, CA, April 1995.

Jarvis, Alka, and Vern Crandall. *Inroads to Software Quality*. Englewood Cliffs, NJ: Prentice Hall, 1997.

Kehoe, Ray, and Alka S. Jarvis. *ISO 9000-3—A Tool for Software Product and Process Improvement*. New York: Springer Verlag, 1995.

Kempf, Mark. *The TL 9000 Guide for Auditors*. Milwaukee: ASQ Quality Press, 2001.

Liebesman, Sandford. "TL 9000—An Update on the Metrics Process." ASQ *Quality Progress* 33, no. 8 (August 2000): 100–102.

———. "TL 9000 Quality Management System Requirements and Metrics." *Continuity*, ASQ Electronics and Communications Division (summer 2000): 6–9.

———. "The Challenges of Auditing TL 9000 Requirements." ASQ *Quality Progress* (March 2001): 106–8.

———. "TL 9000: The Telecommunications Quality Management System." Chap. 60 in *The ASQ ISO 9000:2000 Handbook*. Edited by Charles A. Cianfrani, Joseph J. Tsiakals, and John E. (Jack) West. Milwaukee: ASQ Quality Press, 2002.

Malec, Henry. "TL 9000 Database Repository and Metrics." *The Informed Outlook* 4, no. 6 (June 1999): 4.

Obolewicz, Steve. "Getting to Know TL 9000." *Quality System Update* 9, no. 6 (June 1999): 16–17.

"Participation in the TL 9000 Pilot Program: An Inside View." *The Informed Outlook* 4, no. 10 (October 1999): 3–6.

The QuEST Forum. *TL 9000 Quality System Requirements*. Book One, Release 2.5. Milwaukee: ASQ Quality Press, 1999.

The QuEST Forum. *TL 9000 Quality System Metrics*. Book Two, Release 2.5. Milwaukee: ASQ Quality Press, 1999.

The QuEST Forum. *TL 9000 Quality Management System Requirements Handbook*. Release 3.0. Milwaukee: ASQ Quality Press, 2001.

The QuEST Forum. *TL 9000 Quality Management System Measurements Handbook*. Release 3.0. Milwaukee: ASQ Quality Press, 2001.

"TL 9000 = ISO 9001 + 83 Telecom Sector Requirements + Metrics + Dedicated Data Base." *ISO 9000 News* 8, no. 5 (September/October 1999): 28, 29, 31.

"TL 9000 Metrics: The Key to Continuous Improvement." *The Informed Outlook* 4, no. 5 (May 1999): 24, 25.

"TL 9000 Pilot Program Case Studies Enhance Focus." *The Informed Outlook* 5, no. 4 (April 2000): 1, 27–33.

Watts, Richard, Edward Barabas, and James Gerard. *TL 9000 Implementation Guide*. New York: McGraw-Hill, 2000.

Welch, Steve. "21 Voices for the 21st Century." ASQ *Quality Progress* 33, no. 1 (January 2000): 31–32.

Index

Note: Italicized page numbers indicate illustrations or boxed text.

A

L

M

N

O

S

T

U

V

W